Stochastic Processes and Related Topics

ACADEMIC PRESS RAPID MANUSCRIPT REPRODUCTION

Stochastic Processes and Related Topics

Volume 1 of the Proceedings of the
Summer Research Institute on
Statistical Inference for
Stochastic Processes

Bloomington, Indiana
July 31 – August 9, 1975

EDITED BY

Madan Lal Puri
Professor of Mathematics
Indiana University
Bloomington

ACADEMIC PRESS, INC.   NEW YORK SAN FRANCISCO LONDON 1975

A Subsidiary of Harcourt Brace Jovanovich, Publishers

COPYRIGHT © 1975, BY ACADEMIC PRESS, INC.
ALL RIGHTS RESERVED.
NO PART OF THIS PUBLICATION MAY BE REPRODUCED OR
TRANSMITTED IN ANY FORM OR BY ANY MEANS, ELECTRONIC
OR MECHANICAL, INCLUDING PHOTOCOPY, RECORDING, OR ANY
INFORMATION STORAGE AND RETRIEVAL SYSTEM, WITHOUT
PERMISSION IN WRITING FROM THE PUBLISHER.

ACADEMIC PRESS, INC.
111 Fifth Avenue, New York, New York 10003

United Kingdom Edition published by
ACADEMIC PRESS, INC. (LONDON) LTD.
24/28 Oval Road, London NW1

Library of Congress Cataloging in Publication Data

Summer Research Institute on Statistical Inference
 for Stochastic Processes, Indiana University, 1974.
 Proceedings of the Summer Research Institute on
Statistical Inference for Stochastic Processes,
Bloomington, Indiana, July 31-August 9, 1974.
 "Sponsored by the Institute of Mathematical
Statistics and Indiana University."
 Bibliography: v. 1, p.   ; v. 2, p.
 Includes index.
 CONTENTS: v. 1.   Stochastic processes and related
topics.–v. 2.   Statistical inference and related topics.
 1.   Stochastic processes–Congresses.
2.   Mathematical statistics–Congresses.   I.   Puri,
Madan Lal.   II.   Institute of Mathematical
Statistics.   III.   Indiana. University.
QA274.S964 1975   519.2   74-27522
ISBN 0–12–568001–5 (v. 1)

PRINTED IN THE UNITED STATES OF AMERICA

To Jerzy Neyman

Contents

LIST OF CONTRIBUTORS	viii
FOREWORD	ix
PREFACE	xi
CONTENTS OF VOLUME 2	xiii

J. L. DOOB: *Analytic Sets and Stochastic Processes*	1
L. LECAM: *On Local and Global Properties in the Theory of Asymptotic Normality of Experiments*	13
DAVID R. BRILLINGER: *Statistical Inference for Stationary Point Processes*	55
M. R. LEADBETTER: *Aspects of Extreme Value Theory for Stationary Processes – A Survey*	101
WILLIAM E. PRUITT: *Some Dimension Results for Processes with Independent Increments*	133
DONALD L. IGLEHART: *Conditioned Limit Theorems for Random Walks*	167
G. KALLIANPUR: *Canonical Representations of Equivalent Gaussian Processes*	195
PETER NEY: *A Comparison Method for Critical Branching Processes and an Application to Age Structure*	223
K. B. ATHREYA: *Stochastic Iteration of Stable Processes*	239
J. B. H. KEMPERMAN: *The Ergordic Behavior of a Class of Real Transformations*	249
I. ZHURBENKO: *Mixing Conditions and Asymptotic Theory of Stationary Time Series*	259
WALTER A. ROSENKRANTZ: *Martingales and Diffusion Processes Satisfying Boundary Conditions*	267
VICTOR GOODMAN: *Random Fields Generated by Quantum Fields*	283
D. KÖLZOW: *A Survey of Abstract Wiener Spaces*	293

Contributors

Numbers in parentheses indicate the pages on which the authors' contributions begin.

K. B. Athreya, Department of Mathematics, University of Wisconsin,
Madison, Wisconsin, (239)

David R. Brillinger, Department of Statistics, University of California,
Berkeley, California, (55)

J. L. Doob, Department of Mathematics, University of Illinois,
Urbana, Illinois, (1)

Victor Goodman, Department of Mathematics, Indiana University,
Bloomington, Indiana, (283)

Donald L. L. Iglehart, Department of Operations Research, Stanford University,
Stanford, California, (157)

G. Kallianpur, School of Mathematics, University of Minnesota,
Minneapolis, Minnesota, (195)

J. H. B. Kemperman, Department of Mathematics, The University of Rochester,
Rochester, New York, (249)

D. Kölzow, Mathematisches Institut, Universität Erlangen-Nürnberg,
Erlangen, West Germany, (293)

M. R. Leadbetter, Department of Statistics, University of North Carolina,
Chapel Hill, North Carolina, (101)

L. LeCam, Department of Statistics, University of California,
Berkeley, California, (13)

Peter E. Ney, Department of Mathematics, University of Wisconsin,
Madison, Wisconsin, (223)

William Pruitt, School of Mathematics, University of Minnesota,
Minneapolis, Minnesota, (133)

Walter A. Rosenkrantz, Department of Mathematics and Statistics,
University of Massachusetts, Amherst, Massachusetts, (267)

I. Zhurbenko, Statistical Laboratory, Moscow State University,
Moscow, U. S. S. R., (259)

Foreword

At the beginning of the 1974 Summer Research Institute, the participants decided, by common consent, that the sessions would be dedicated to one of their colleagues, Professor Jerzy Neyman.

The enthusiasm with which the decision was made will not surprise anyone who has even a moderate acquaintance with our domain. However, the constantly expanding scope of Neyman's activities, his overwhelming presence, and his half century of achievements makes it difficult to give a written description of the reasons for this dedication. We shall discuss here only a few aspects which, although they are common knowledge, can still be restated with some benefit.

Contrary to what one might be tempted to guess from the sophistication and variety of the articles in these proceedings, our understanding of stochastic stochastic processes is not very old. An inkling of their difficult history can be gathered from the fact that as recently as 1947 J.L. Doob felt it necessary to argue that they are respectable mathematical objects which can be studied through measure theoretical methods. Time series were of course more easily acceptable to the mathematicians, but their statistical study was another matter. In his 1948 essay entitled "Cybernetics" Norbert Wiener writes,

> The development of our theory beyond this point, as a practical statistical theory, involves an extension of existing methods of sampling. The author and others have made a beginning in this direction. It involves all the complexities of the use, either of Bayes' Law on the one hand, or of those terminological tricks in the theory of likelihood, on the other, which seem to avoid the necessity for the use of Bayes law, but which in reality transfer the responsibility for its use to the working statistician, or the person who ultimately employs his results.

This passage, which, coming from N. Wiener, is a bit surprising, has been quoted here to point out that even at that time, principles of statistical inference were neither widely known nor easily understood. One reason for this is that the principles themselves, or, more exactly, principles that did not involve "terminological tricks" or outright dogma, had been clearly formulated only about 40 years ago. Furthermore, at the time, they were very badly received by the established statistical community.

With the benefit of hindsight, one can now find traces of similar ideas in the works of Laplace and Edgeworth, for instance. However, at the time Neyman and Pearson stated the principles on which we now base all of statistical decision theory they must have appeared as subversive as Copernicus suggesting that the Earth is not the center of the Universe. Many statisticians, including the great R.A.

Fisher, fought bitterly against the idea that one can describe probabilistically the performance characteristics of a test or estimation procedure and that this description is often relevant to the choice of the procedure. The idea was too novel to be readily acceptable in connection with the theory of testing hypotheses. Neyman had even more difficulty in getting it accepted in connection with the theory of confidence intervals. However, it eventually became the basis for A. Wald's theory of statistical decision functions and gained widespread use, if not acceptance, under that form.

The ideas and techniques issued from these principles did not penetrate deeply into the analysis of stochastic processes until the early 1950s. The reason is simply that their application presumes a certain amount of work which, as Neyman himself has repeatedly emphasized, constitutes one of the main and most difficult tasks of the statistician, namely the construction of adequate stochastic models. One of the characteristics of Neyman's attack on a practical problem is that he will spend an unlimited amount of energy studying the physical or biological mechanism of the phenomena under scrutiny. Then, keeping what appears essential in the mechanism, he will formulate a stochastic model describing the situation and tailor whatever statistical methods are needed to that model. Examples of this approach can be found in his introduction of "contagious" distributions, his study of accident proneness, of clustering of galaxies, of the mechanism of carcinogenesis and many other studies. It is an interesting exercise to see how the papers collected here are in various degrees permeated by these ideas.

Having been a student of J. Neyman for about 25 years, I feel very inadequate in expressing my respect for him as a scholar, a teacher, and an exceptional human being. However I am very honored to have been a participant in a conference so fittingly dedicated to him.

The success of the conference itself is due to Professor Madan L. Puri whose efforts spanned the terms of at least three Presidents of the Institute of Mathematical Statistics, namely, Professors Bose, Bahadur, and myself. I am very grateful to Professor Puri for giving me the opportunity of adding these few words of introduction to the proceedings of the meeting.

L. LE CAM

Preface

The past several years have seen the creation and extension of a very inclusive theory of statistical inference for stochastic processes. Many of the research workers who have been concerned with both probability and statistics felt the need for a meeting that would provide an opportunity for personal contacts among scholars whose fields of specialization cover broad spectra in both these areas, discuss major open problems, and provide stimulation for further research through the lectures of carefully selected scholars, and introduce to promising graduate students the latest research techniques and stimulate their interest in research. To meet these purposes, the Summer Research Institute on Statistical Inference for Stochastic Processes was held at Indiana University, Bloomington, from July 31 to August 9, 1974. The Institute was sponsored by the Institute of Mathematical Statistics and Indiana University. Professors Ward B. Schaap, acting Vice-Chancellor for Administration and Budgetary Planning at Indiana University, Bloomington, and Jerzy Neyman, University of California, Berkeley were very kind to make the opening remarks, and Dr. B.R. Agins, National Science Foundation, welcomed the guests. Dean R.C. Buck, Professors Lucien Le Cam and Jacob Wolfowitz were the main speakers at the banquet.

The Summer Research Institute Committee consisted of Professors R.E. Bechhofer, Z.W. Birnbaum, H.T. David, Wassily Hoeffding, A.W. Marshall, Herman Rubin, Milton Sobel, and myself. I take this opportunity to express my sincere thanks to the members of this committee for the overall help they gave me in making the Summer Research Institute a great success. In addition I also benefited greatly from the help and encouragement I received from Professors Shanti S. Gupta, Jack Kiefer, Lucien LeCam, Malcolm Ross Leadbetter, Erich Lehmann, Emanuel Parzen, Ronald Pyke, and Murray Rosenblatt among others. It is a pleasure to express my deep sense of appreciation to all of them.

The proceedings of the Summer Research Institute are divided into two parts. The contents of both these parts are given in each volume.

Professors John Beekman, Peter J. Bickel, John J. Birch, Saul Blumenthal, David Brillinger, John Chadam, H.T. David, J.L. Doob, Robert Glassey, Shanti S. Gupta, Paul Halmos, Leo Katz, J.H.B. Kemperman, D. Kölzow, Lucien Le Cam, Charles Newman, Jerzy Neyman, Emanuel Parzen, Ronald Pyke, A.T. Bharucha-Reid, Murray Rosenblatt, Herman Rubin, Malcolm Ross Leadbetter, Malcolm Quine, Stephen M. Samuels, Moin Siddiqui, Nicholas Spulber, Jagdish N. Srivastava, James Stapleton, Jacob Wolfowitz and Drs. B.R. Agins, P.R. Krishnaiah and V.R. Uppuluri presided over different sessions. Professors Emanuel Parzen and Ronald Pyke held some special sessions for the benefit of younger participants.

It is also a pleasure to express my appreciation to John Chadam, Victor Goodman, S.K. Mitra, J.S. Rao and Debbie Wilson for taking care of problems connected

PREFACE

with local arrangements. I would also like to thank the contributors for the fine spirit of cooperation and the prompt handling of the correspondence.

The Summer Research Institute would not have been possible without the very generous support from the National Science Foundation. My heartiest thanks are due to Drs. William H. Pell and B.R. Agins who took personal interest in the project, and gave me all the help I needed in making the Institute successful. It is also a pleasure to express my deep sense of appreciation to Indiana University for providing excellent facilities to hold this conference; in particular to Dr. Harrison Shull, Vice-President, Research and Development who, in spite of his multifarious activities, took personal interest in the project and gave me very valuable assistance, advice and the financial support from his office.

The editorial work of the project of this magnitude could not have been accomplished without the aid of my distinguished friend and colleague, Victor Goodman, who assumed full responsibility while I was in Göttingen on a Senior U.S. Scientist Award. Victor worked with utmost care, rare dedication, and tireless persistence and spent a considerable amount of time from the initiation to the completion of this project. It is a pleasure to express here my deep appreciation.

Last but not least my thanks go to the personnel of the Academic Press for taking care of the publication of the proceedings with great patience.

MADAN LAL PURI

Contents of Volume 2

J. PFANZAGL: *On Asymptotically Complete Classes*	1
RONALD PYKE: *Multidimensional Empirical Processes: Some Comments*	45
JON A. WELLNER: *Monte Carlo of Two-Dimensional Brownian Sheets*	59
KEH-SHIN LII and MURRAY ROSENBLATT: *Asymptotic Results on a Spline Estimate of a Probability Density*	77
RICHARD A. VITALE: *A Bernstein Polynomial Approach to Density Function Estimation*	87
SHANTI S. GUPTA and DENG-YUAN HUANG: *On Some Parametric and Nonparametric Sequential Subset Selection Procedures*	101
PRANAB KUMAR SEN: *Rank Statistics, Martingales and Limit Theorems*	129
Y. S. CHOW and K. K. LAN: *Optimal Stopping Rules for X_n/n and S_n/n*	159
MICHAEL J. KLASS: *A Survey of the S_n/n Problem*	179
HERMAN RUBIN: *Some Non-Standard Problems of Inference in Stochastic Processes*	203
GEORGE G. ROUSSAS: *Asymptotic Properties of Maximum Probability Estimates in the IID Case*	211
PREM S. PURI: *A Stochastic Process Under the Influence of Another Arising in the Theory of Epidemics*	235
J. S. RUSTAGI: *Inference in a Distribution Related to a 2 × 2 Markov Chain*	257
C. B. BELL: *Statistical Inference for Some Special Families of Stochastic Processes*	273
RONALD BIONDINI and M. M. SIDDIQUI: *Record Values in Markov Sequences*	291

Analytic Sets and Stochastic Processes

by J. L. Doob
University of Illinois

The theory of analytic sets was initiated in 1917 but the significance of the theory for probability was not realized until the publication of Hunt's 1957 papers on Markov processes. What is presented here is an introductory outline of part of the theory in its present form, with a typical application to stochastic process theory. A few proofs are included, to give the flavor of the subject.

1. Notation and conventions. If $\mathcal{Y}$ is any collection of subsets of a space, $\mathcal{Y}_\sigma$ [$\mathcal{Y}_\delta$] denotes the class of countable unions [intersections] of sets in $\mathcal{Y}$. In particular, if $\mathcal{Y}_\sigma = \mathcal{Y}_\delta = \mathcal{Y}$ and if $\mathcal{Y}$ contains the complements of its members, $\mathcal{Y}$ is a σ algebra.

A space Y together with a collection $\mathcal{Y}$ of its subsets, including the empty set, is called a paved space, or a measurable space if $\mathcal{Y}$ is a σ algebra. If $(X,\mathcal{X})$ and $(Y,\mathcal{Y})$ are paved spaces, the product paved space is defined as $(X \times Y, \mathcal{X} \times \mathcal{Y})$, where $\mathcal{X} \times \mathcal{Y}$, the product paving, is defined as $\{X_0 \times Y_0 : X_0 \in \mathcal{X}, Y_0 \in \mathcal{Y}\}$. If the factor paved spaces are measurable spaces the product measurable space is defined as $\{X \times Y, \mathcal{X} \otimes \mathcal{Y}\}$, where $\mathcal{X} \otimes \mathcal{Y}$, the product σ algebra, is the σ algebra generated by $\mathcal{X} \times \mathcal{Y}$. We shall use the fact that $\mathcal{X} \otimes \mathcal{Y}$ is contained in any class Γ of subsets of

X×Y for which $\Gamma \supset \chi \times \psi$ and $\Gamma = \Gamma_\sigma = \Gamma_\delta$. A function f from X into Y is called measurable if $f^{-1}(\psi) \subset \chi$.

Let X be a topological space. The class of Borel subsets of X is the σ algebra generated by the closed sets. A G_δ subset of X is defined as a countable intersection of open sets. If X is metric, a family of sets containing the closed sets contains the Borel sets if $\Gamma = \Gamma_\sigma = \Gamma_\delta$.

2. Topological context. The following standard topological facts provide the context for the theory of analytic sets in metric spaces to be discussed in Section 8.

(a) A G_δ of a complete metric space is homeomorphic with a complete metric space.

(b) Any 1-1 bicontinuous map of a complete metric space S into a metric space takes S into a G_δ.

(c) A complete separable metric space is homeomorphic with a G_δ in the interval [0,1].

Combining (a) and (c) a G_δ of a complete separable metric space is homeomorphic with a G_δ in the interval [0,1].

A Polish space is defined as a Hausdorff space homeomorphic with a complete separable metric space. A G_δ in a Polish space is itself a Polish space. The usual state space of Markov process theory is locally compact and second countable. Such a space is Polish because it is an open subset of its one point compactification and the latter space is compact and metrizable. If X and Y are Polish, their product space in the product topology is Polish.

3. The Baire null space. This space is the

metric space of sequences $n_\cdot = (n_1, n_2, \ldots)$ of strictly positive integers, with

$$\text{dist.}(m_\cdot, n_\cdot) = [\min j: m_j \neq n_j]^{-1} \text{ if } m_\cdot \neq n_\cdot.$$

The space is complete and separable. The continued fraction representation

$$x = \cfrac{1}{n_1 + \cfrac{1}{n_2 + \cfrac{1}{n_3 + \cdots}}}$$

valid and unique for x irrational in [0,1] is a homeomorphism (illustrating Section 2(c)) between the Baire null space and the set of irrationals in [0,1], a G_δ in this interval.

Every Polish space Y is the continuous image of the Baire null space by means of the following map, in which (without loss of generality) Y is assumed complete separable metric. Choose closed nonempty sets $Y_1, Y_2, \ldots$ of diameters <1, with union Y. If $Y_{n_1 \ldots n_k}$ has been chosen, choose closed nonempty sets $Y_{n_1 \ldots n_k 1}, Y_{n_1 \ldots n_k 2}, \ldots$ of diameters <1/(k+1), with union $Y_{n_1 \ldots n_k}$. The desired map is that taking $n_\cdot$ into the single point in $Y_{n_1} \cap Y_{n_1 n_2} \cap \cdots$.

4. Suslin schemes. Let $(Y, \mathcal{Y})$ be a paved space. A Suslin scheme, as defined by Suslin in 1917, is a map $Y_\cdot$ from finite sequences of strictly positive integers into $\mathcal{Y}$:

$$(n_1,\ldots,n_k) \Rightarrow Y_{n_1\ldots n_k} \in \mathcal{Y}.$$

To each infinite sequence $n_\bullet$ corresponds the intersection $Y_{n_1} \cap Y_{n_1 n_2} \cap \ldots$. The uncountable union $\cup_{n_\bullet} Y_{n_1} \cap Y_{n_1 n_2} \cap \ldots$ of all these intersections is called the nucleus of the Suslin scheme and is said to be analytic over $\mathcal{Y}$. The collection of these nuclei will be denoted by $G(\mathcal{Y})$.

If $A \in \mathcal{Y}$ and if $Y_{n_1\ldots n_k} = A$ for all k and $n_1, \ldots, n_k$ then A is the nucleus of $Y_\bullet$ and therefore $\mathcal{Y} \subset G(\mathcal{Y})$. Less trivially, $G(\mathcal{Y})_\sigma = G(\mathcal{Y})$. To see this suppose that $A_j \in G(\mathcal{Y})$, say

$$A_j = \cup_{n_\bullet} Y_{jn_1} \cap Y_{jn_1 n_2} \cap \ldots, \quad Y_{jn_1\ldots n_k} \in \mathcal{Y}.$$

Let α be a 1-1 map from the set of strictly positive integers into the set of pairs of these integers and define

$$Z_{m_1\ldots m_k} = Y_{jn_1 m_2\ldots m_k}, \quad \text{where } (j,n_1) = \alpha(m_1).$$

Then $\cup_j A_j$ is the nucleus of $Z_\bullet$. A slightly more complicated argument shows that $G(\mathcal{Y})_\delta = G(\mathcal{Y})$. A still more complicated argument shows that $G[G(\mathcal{Y})] = G(\mathcal{Y})$, but we omit this, as well as a much more perspicuous proof of this result based on the characterization of $G(\mathcal{Y})$ in Section 7.

5. Theorem. _If_ $(X,\mathcal{X})$ _and_ $(Y,\mathcal{Y})$ _are paved spaces and if f is a function from X into Y with_ $f^{-1}(\mathcal{Y}) \subset \mathcal{X}$ _then_ $f^{-1}[G(\mathcal{Y})] \subset G(\mathcal{X})$.

This theorem is a trivial consequence of the fact that the inverse image under f of an arbitrary intersection [union] of sets is the intersection [union] of their inverse images. In the most common application the paved spaces are measurable spaces, so that f is by hypothesis a measurable transformation. Thus every σ algebra $\mathcal{Y}$ has an extension $G(\mathcal{Y})$ with the property that, in intuitive language, the measurability relationship is preserved in the extension.

6. Universally measurable sets. The extension from $\mathcal{Y}$ to $G(\mathcal{Y})$ is not the only useful extension that preserves the measurability relationship in the sense just described. Let $(Y,\mathcal{Y})$ be a measurable space and let μ be a σ finite measure on $\mathcal{Y}$. The class $\mathcal{Y}_\mu$ of μ measurable sets, obtained by completing μ, is the σ algebra generated by $\mathcal{Y}$ and the class of subsets of μ null sets. The intersection $\mathcal{U}(\mathcal{Y}) = \bigcap_\mu \mathcal{Y}_\mu$ is a σ algebra, the class of 'universally measurable' sets (over $\mathcal{Y}$). Theorem 5 remains true if in its statement $G(\mathcal{X})$ and $G(\mathcal{Y})$ are replaced by $\mathcal{U}(\mathcal{X})$ and $\mathcal{U}(\mathcal{Y})$. According to Lusin's theorem (Section 10) $G(\mathcal{Y}) \subset \mathcal{U}(\mathcal{Y})$. Every σ finite measure μ on $\mathcal{Y}$ has a unique measure extension to $\mathcal{U}(\mathcal{Y})$, the restriction to $\mathcal{U}(\mathcal{Y})$ of the completion of μ.

7. The following characterization of $G(\mathcal{Y})$ is important for the applications to stochastic process theory.

Theorem. <u>Let $(Y, \mathcal{Y})$ be a paved space and let $(X, \mathcal{X})$ be a topological space together with its class of compact sets. Then the projection on Y of a set in $G(\mathcal{X} \times \mathcal{Y})$ is in $G(\mathcal{Y})$.</u> For a suitable choice of $(X, \mathcal{X})$, <u>in which X can be taken compact metric and $\mathcal{X}$ the class of compact subsets of X, $G(\mathcal{Y})$ is the class of projections on Y of the sets in $(\mathcal{X} \times \mathcal{Y})_{\sigma\delta}$.</u>

The first assertion (which is all we prove) is true because if $\hat{A} \in G(\mathcal{X} \times \mathcal{Y})$, say

$$(7.1) \quad \hat{A} = \bigcup_{n_{\bullet}} (X_{n_1} \times Y_{n_1}) \cap (X_{n_1 n_2} \times Y_{n_1 n_2}) \cap \ldots$$

$$= \bigcup_{n_{\bullet}} (X_{n_1} \cap X_{n_1 n_2} \cap \ldots) \times (Y_{n_1} \cap Y_{n_1 n_2} \cap \ldots),$$

where $X_{n_1 \ldots n_k} \in \mathcal{X}$ and $Y_{n_1 \ldots n_k} \in \mathcal{Y}$, and if $A_{\bullet}$ is defined by

$$(7.2) \quad A_{n_1 \ldots n_k} = Y_{n_1 \ldots n_k} \text{ if } X_{n_1} \cap \ldots \cap X_{n_1 \ldots n_k} \neq \emptyset$$

$$= \emptyset \text{ otherwise,}$$

then the projection of $\hat{A}$ on Y is the nucleus of $A_{\bullet}$ and is therefore in $G(\mathcal{Y})$.

Corollary. <u>In the first assertion of the theorem if X is locally compact second countable, with class $\mathcal{B}_X$ of Borel sets, and if $(Y, \mathcal{Y})$ is a measurable space, the projection on Y of a set in $G(\mathcal{B}_X \otimes \mathcal{Y})$ is in $G(\mathcal{Y})$.</u>

In view of the theorem we need only show that $G(\mathcal{X} \times \mathcal{Y}) \supset G(\mathcal{B}_X \otimes \mathcal{Y})$. (Inclusion in the other direction is of course trivial.) Since X is metrizable and since $G(\mathfrak{X}) = G(\mathfrak{X})_\sigma = G(\mathfrak{X})_\delta$, $G(\mathfrak{X})$ must contain $\mathfrak{B}_X$ (see Section 1). Hence

$$(7.3) \quad G(\mathfrak{X} \times \mathfrak{Y}) = G[G(\mathfrak{X} \times \mathfrak{Y})] \supset G(\mathfrak{B}_X \times \mathfrak{Y}) \supset \mathfrak{B}_X \otimes \mathfrak{Y},$$

where the last inclusion is true because (see Section 1)

$$(7.4) \quad G(\mathfrak{B}_X \times \mathfrak{Y}) = G(\mathfrak{B}_X \times \mathfrak{Y})_\sigma = G(\mathfrak{B}_X \times \mathfrak{Y})_\delta \supset \mathfrak{B}_X \otimes \mathfrak{Y}.$$

Operating with G on the extreme terms of (7.3) yields the desired inclusion.

8. **Analytic sets.** A set will be called 'analytic' if it is a subset of a metrizable space and if it is analytic over the class of closed sets of the space. The class of analytic subsets of a metrizable space, a class containing countable unions and intersections of its members, must contain the Borel subsets of the space. In general there are analytic sets which are not Borel sets, for example when the space is Euclidian N space.

Theorem. <u>The following conditions on a subset A of a Polish space are equivalent.</u>

(a) <u>A is analytic.</u>

(b) <u>A is the continuous image of a Polish space.</u>

(c) <u>A is the continuous image of a G_δ of a compact metric space.</u>

(d) <u>A is the continuous image of the Baire null space.</u>

(e) <u>A is the continuous image of the set of irrationals in $[0,1]$.</u>

In the older literature an analytic set is sometimes defined as a set satisfying (e). Most of the equivalences of this theorem are trivial consequences of the facts outlined in Sections 2 and 3. We prove only that (d) implies (a). Under (d), $A = f(B)$, where B is the Baire null space, and f is continuous from B into the Polish space containing A. Let $X_{n_1 \ldots n_k}$ be the set of points of B with first k coordinates $n_1, \ldots, n_k$. This set is closed and has diameter $1/(k+1)$. Let $Y_{n_1 \ldots n_k}$ be the closure of $f(X_{n_1 \ldots n_k})$. If $n_. \in B$,

(8.1) $\quad X_{n_1 \ldots n_k} \downarrow \{n_.\}, \quad Y_{n_1 \ldots n_k} \downarrow \{f(n_.)\} \quad (k \to \infty)$

and therefore A is the nucleus of $Y_.$ and so analytic, as was to be proved.

A trivial consequence of these characterizations of analytic sets is that if f is a continuous function from a Polish space X into a Polish space Y it takes analytic sets into analytic sets. It is only slightly less trivial that the conclusion remains true even if f is merely Borel measurable. In fact the graph of f is then a Borel subset of X×Y, therefore analytic. Furthermore if A is an analytic subset of X, A×Y is an analytic subset of X×Y so the intersection $\hat{A}$ of the graph of f with A×Y, that is, the graph of the restriction of f to A, is an analytic subset of X×Y. Finally f(A) is the projection of $\hat{A}$ on Y, a continuous map, so f(A) is an analytic subset of Y.

9. Choquet capacity. Let $(X, \mathcal{I})$ be a paved space for which $\mathcal{I}$ is invariant under finite unions and intersections. A function I from the class of subsets of X into $[-\infty, \infty]$ is called a Choquet capacity (relative to $\mathcal{I}$) if it satisfies the following conditions.

(a) $X_1 \subset X_2$ implies that $I(X_1) \leq I(X_2)$.

(b) $X_n \uparrow X_\infty$ implies that $I(X_n) \to I(X_\infty)$.

(c) $X_n \in \mathcal{I}$ and $X_n \downarrow X_\infty$ implies that $I(X_n) \to I(X_\infty)$.

A subset A of X is said to be I capacitable if I is a Choquet capacity and if $I(A) = \sup \{I(B): B \subset A, B \in \mathcal{I}_\delta\}$.

The fundamental theorem here is the Choquet Capacity Theorem (1955). <u>The sets of $\mathcal{A}(\mathcal{I})$ are I capacitable.</u> (Proof omitted.)

10. Lusin's Theorem (1917). Let $(Y, \mathcal{Y}, \mu)$ be a σ finite measure space and define the outer measure μ^* of a subset A of Y by

$$\mu^*(A) = \inf \{\mu(B): A \subset B \in \mathcal{Y}\}.$$

Let $\mathcal{Y}_\mu$ be the domain of the completion of μ, as in Section 6. Let $X = Y$, let $\mathcal{I}$ be the class of sets in $\mathcal{Y}$ of finite measure and let $I = \mu^*$. Then I is a Choquet capacity relative to $\mathcal{I}$ and capacitability of a set of finite outer measure is μ measurability, that is membership in $\mathcal{Y}_\mu$. The Choquet capacity theorem implies that the sets of $\mathcal{A}(\mathcal{I})$ of finite outer measure and therefore (by σ additivity) all sets of $\mathcal{A}(\mathcal{I})$ are in $\mathcal{Y}_\mu$. Since $\mathcal{A}(\mathcal{I}) = \mathcal{A}(\mathcal{Y})$ we have

$G(\mathcal{y}) \subset \mathcal{y}_\mu$. (Actually $G(\mathcal{y}_\mu) = \mathcal{y}_\mu$ because $\mathcal{y}$ can be replaced by $\mathcal{y}_\mu$ in this discussion.) Since this inclusion holds for all μ the sets of $G(\mathcal{y})$ are universally measurable, $G(\mathcal{y}) \subset u(\mathcal{y})$.

11. Example. Consider the following classes of subsets of the line: the class $\mathcal{B}$ of Borel sets; the class of analytic sets; the class $u(\mathcal{B})$ of universally measurable sets over $\mathcal{B}$; the class of Lebesgue measurable sets. Each of these classes can be shown to be strictly larger than the preceding one.

The projection on a line of a plane Borel set is not necessarily a Borel set but, as the continuous image of an analytic set, is an analytic subset of the line. This projection is therefore Lebesgue measurable in one dimension. The projection on a line of a Lebesgue measurable (two dimensional) set is not necessarily analytic or even Lebesgue measurable in one dimension however because every subset of the line is its own projection and is Lebesgue measurable in two dimensions.

12. Application to stochastic process theory. Let X be a compact real interval and let $\mathcal{X}[\mathcal{B}_X]$ be the class of compact [Borel] subsets of X. Let $(S, \mathcal{S})$ be a locally compact second countable space together with its Borel subsets. Let $\{x(t), t \in X\}$ be a stochastic process with state space S, defined on the probability measure space $\{\Omega, \mathcal{F}, P\}$. It is supposed that P is complete on $\mathcal{F}$. The random variable $x(t)$ has value $x(t, \omega)$ at ω in Ω. If $A \subset S$ define

$$\hat{A} = \{(t, \omega): \quad x(t, \omega) \in A\},$$

$$\widetilde{A} = \{\omega: x(t,\omega) \in A \text{ for some } t \text{ in } X\}.$$

The set $\widetilde{A}$ is the projection of $\hat{A}$ on Ω and is the set of those ω yielding sample paths $x(.,\omega)$ which hit A.

Consider the following problem. For which sets A is $P\{\widetilde{A}\}$ well defined? That is, for which sets A is $\widetilde{A}$ in $\mathfrak{F}$, so that the probability that a sample path hits A is well defined? We suppose from now on that the map $(t,\omega) \Rightarrow x(t,\omega)$ from $(X \times \Omega, \mathcal{B}_X \otimes \mathfrak{F})$ into $(S, \mathcal{S})$ is measurable. That is, $A \in \mathcal{S}$ implies that $\hat{A} \in \mathcal{B}_X \otimes \mathfrak{F}$. This measurability hypothesis is true, for example, if the process paths are right continuous. According to Theorem 5, $A \in \mathcal{G}(\mathcal{S})$ implies that $\hat{A} \in \mathcal{G}(\mathcal{B}_X \otimes \mathfrak{F})$, and (Corollary 7 and Lusin's theorem) $\widetilde{A} \in \mathfrak{F}$. Thus the probability that a sample path hits an analytic subset of S is well defined.

It is instructive to deal with this problem from the point of view of Choquet capacities. With commonly used hypotheses on the stochastic process it can be shown that if A is a compact subset of S then $\widetilde{A} \in \mathfrak{F}$. Define $I(A) = P(\widetilde{A})$ for A a compact subset of S. Extend I to the class of open subsets of S by

$$I(A) = \sup \{I(B): B \text{ compact}, B \subset A\} \quad (A \text{ open})$$

and define I for an arbitrary subset A of S by

$$I(A) = \inf \{I(B): A \subset B, B \text{ open}\}.$$

Under suitable hypotheses on the stochastic process it is shown that this definition of I is selfconsistent and that I is a Choquet capacity relative to the class of compact subsets of S. By the definitions, $I(A)$ is the probability that a process sample path hits a point of A if A is either compact or open. The definition of capacitability implies that if A is capacitable for I, in particular if A is analytic, then $\widetilde{A} \in \mathfrak{J}$ and $P(\widetilde{A}) = I(A)$ is the probability that a process path hits a point of A. Moreover if A is capacitable there is an open superset A_1 of A and a compact subset A_0 of A such that $P(\widetilde{A}_1)$ and $P(\widetilde{A}_0)$ are arbitrarily close to $P(\widetilde{A})$.

REFERENCES FOR FURTHER READING

Claude Dellacherie: Ensembles Analytiques, Capacités. Mesures de Hausdorff. Lecture Notes in Mathematics 295. Springer Verlag (1972).; Capacités et Processus Stochastiques. Ergebnisse der Mathematik und ihrer Grenzgebiete 67. Springer Verlag (1972).

Felix Hausdorff: Set Theory. Chelsea Publishing Co. (1962).

Kazimierz Kuratowski: Topology. Academic Press (1966).

Paul A. Meyer: Probability and Potentials. Blaisdell Publishing Co. (1966).

ON LOCAL AND GLOBAL PROPERTIES IN THE THEORY OF ASYMPTOTIC NORMALITY OF EXPERIMENTS[*]

by L. Le Cam

University of California, Berkeley

1. <u>Introduction</u>. Let Θ be an arbitrary set. It will be assumed that an experiment indexed by Θ is given by a σ-field Q and a map $\mathcal{E} = \{P_\theta : \theta \in \Theta\}$ from Θ to the space of probability measures on Q.

The idea that "when the number of observations is large" such a family can be approximated by a Gaussian family of distributions occurs in the remarkable paper [1] of A. Wald. Subsequently the present author suggested that asymptotically sufficient estimates providing the kind of approximation described by Wald can often be obtained by a two steps "adaptive" procedure as follows. One first finds a good but rough estimate $\hat{\theta}$ of the value of θ. Then, in the vicinity of the estimated value, one approximates the logarithms of likelihood ratios by a suitable quadratic expression. One proceeds as if the quadratic expression was obtained from the logarithms of likelihood ratios of an actual Gaussian family of measures.

---

[*] This research was supported by National Science Foundation Grant GP-31091X.

Intuition suggests that the procedure will work in a large variety of different situations. A proof to this effect was first given in a very special context in [2]. Later on generalizations were described in a series of papers. (See [3] for a summary and references.) In most of these generalizations the existence of a suitable preliminary estimate is not proved, but assumed outright, as explained for instance in Section 12 of [3].

In various attempts to formulate general statements concerning the existence of these estimates the present author found himself repeatedly led to use restrictions on the dimensionality of the parameter set Θ. Similar restrictions arise also naturally in the passage from local to global necessary to demonstrate the validity of the two step procedure. Although some restrictions are unavoidable, one cannot claim that the dimensionality restrictions are actually necessary. It is rather easy to find perfectly honorable examples where the assumptions do not hold but estimates exist and are usable. However, since no other general conditions have yet been formulated we present here a few results which are proved under dimensionality restrictions.

Specifically, the contents of the paper are as follows.

In Section 2 we consider an experiment $\mathcal{E} = \{P_\theta; \theta \in \Theta\}$ and describe a construction of confidence sets. The description is given in some detail, since a previous description, published in

[4], incorporates a disabling and well-known fallacy.

Section 3 considers two experiments $\mathcal{E} = \{P_\theta; \theta \in \Theta\}$ and $\mathcal{J} = \{Q_\theta; \theta \in \Theta\}$. For any subset $S \subset \Theta$ let $\mathcal{E}_S = \{P_\theta; \theta \in S\}$ and let $\mathcal{J}_S = \{Q_\theta; \theta \in S\}$. One assumes that for "small" sets S the deficiencies $\delta(\mathcal{E}_S, \mathcal{J}_S)$ are small and that $\mathcal{E}$ provides good estimates of θ. The result is that, under suitable dimensionality restrictions, one can approximate the whole experiment $\mathcal{J}$ through a post-experimental randomization of $\mathcal{E}$.

Section 4 applies the scheme of Section 2 to a situation where the experiment $\mathcal{E}$ is made up of many independent observations. An improvement available when the observations are not only independent but identically distributed is given in Section 5.

Section 6 returns to the general independent case and the possibility of approximations by Poisson processes.

Section 7 is devoted to the local behavior of the experiments. It recalls a general description of approximability by Gaussian experiments and indicates briefly how the conditions can be verified in the case of independent observations. For the dependent cases, mention is made of the possibility of applying the theorems of A. Dvoretzky [5] and D. L. McLeish [6].

2. <u>A construction scheme</u>. Let Θ be a set and let $\mathcal{G}$ be a σ-field. For each θ, let P_θ be a probability measure on $\mathcal{G}$. Let W be a real-valued function defined on $\Theta \times \Theta$. We shall proceed below as if W was a metric on Θ, but the only relation

actually needed here is the symmetry property $W(s,t) = W(t,s)$.

A subset A of Θ has a diameter $\text{diam}(A) = \sup\{W(s,t); s \in A, t \in A\}$. Let $\{b_\nu; \nu = 0, 1, 2, \ldots, \}$ be a sequence of numbers such that $b_\nu > b_{\nu+1} > 0$ for all ν. Let $\{a_\nu\}$ be another sequence such that $0 \leq a_\nu < b_\nu$. For each ν, let $\{A_{\nu,i}; i \in I_\nu\}$ be a partition of Θ by sets $A_{\nu,i}$ whose diameter does not exceed a_ν.

The construction will be described assuming that the following conditions hold.

Assumption A1. *The diameter of Θ does not exceed b_0. The partition $\{A_{\nu,i}; i \in I_\nu\}$ has minimum cardinality among all those subject to the restriction that $\text{diam } A_{\nu,i} \leq a_\nu$. Furthermore, I_ν is a finite set.*

One can always insure the feasibility of such a restriction by imposing the following requirement.

Assumption A2. *The set Θ is finite.*

Passage from the finite set situation to the general one is described at the end of Section 3.

Under Assumptions 1 and 2 one can proceed as follows. Call a pair (i,j) of elements of I_ν a ν-distant pair if there are elements $s \in A_{\nu,i}$ and $t \in A_{\nu,j}$ such that $W(s,t) \geq b_\nu$. For such a pair let $\varphi_{\nu,i,j}$ be a test of $A_{\nu,j}$ against $A_{\nu,i}$. Assume that the tests are selected so that

$$\varphi^2_{\nu,i,j} = \varphi_{\nu,i,j} = 1 - \varphi_{\nu,j,i}.$$

Let $B_0 = \Theta$. If $B_0, B_1, \ldots, B_{\nu-1}$ have been constructed, let $J_\nu \subset I_\nu$ be the set of indices

$i \in I_\nu$ such that $A_{\nu,i}$ intersects $B_{\nu-1}$.
Furthermore, if $i \in J_\nu$, let $J_\nu(i)$ be the set of $j \in J_\nu$ which are ν-distant from i. Define $\psi_{\nu,i}$ by

$$\psi_{\nu,i} = \inf_j \{\varphi_{\nu,i,j}; \ j \in J_\nu(i)\}.$$

For each sample point x there is a certain set $S_\nu(x)$ of indices $i \in J_\nu$ for which $\psi_{\nu,i}(x) = 1$. Let B_ν be the set

$$B_\nu = \bigcup_i \{A_{\nu,i} \cap B_{\nu-1}; \ i \in S_\nu(x)\}.$$

Note that by construction $B_\nu \subset B_{\nu-1}$. Also diam $B_\nu < b_\nu$, since each $A_{\nu,i}$ has a diameter at most $a_\nu < b_\nu$ and since for any ν-distant pair (i,j) one has $\psi_{\nu,i} \psi_{\nu,j} = 0$.

The construction just described is the one we meant to use in [4]. However, the paper in question would allow arbitrary choices of the covers $\{A_{\nu,i}\}$ after $B_{\nu-1}$ has been obtained. This invalidates the computation of covering probabilities given in [4] and reproduced on page 81 of [3].

To obtain an indication on the probabilistic performance of the construction we shall use the following notation. For any two sets A_i, $i = 1,2$, $A_i \subset \Theta$ let $\pi(A_1, A_2)$ be the number

$$\pi(A_1, A_2) = \inf_\varphi \sup_{s,t} \{\int (1-\varphi) dP_s + \int \varphi dP_t\}$$

where the supremum is taken for $s \in A_1$, $t \in A_2$ and where the infimum is taken over all measurable φ

such that $0 \leq \varphi \leq 1$. For any pair $(A_{\nu,i}, A_{\nu,j})$ one can find tests $\varphi_{\nu,i,j} = \varphi^2_{\nu,i,j}$ such that

$$2\pi[A_{\nu,i}, A_{\nu,j}] \geq \sup_{s,t}\{\int(1-\varphi_{\nu,i,j})dP_s + \int \varphi_{\nu,i,j}dP_t\}.$$

(Here s varies in $A_{\nu,i}$ and t in $A_{\nu,j}$.)
We shall assume that the tests used in the construction do satisfy this inequality.

For each $\theta \in \Theta$ and each ν, let $U_\nu(\theta)$ be the set of points t such that there is an $s \in \Theta$ for which $W(\theta,s) < b_{\nu-1}$ and $W(s,t) \leq a_\nu$.

Assuming that the partitions and tests are selected as described we can assert the following.

Proposition 1. Let Assumptions 1 and 2 hold. Suppose that there is a number $\beta(\nu)$ such that $\pi(A_{\nu,i}, A_{\nu,j}) \leq \beta(\nu)$ for all ν-distant pairs (i,j). Suppose also that there is a number $C(\nu)$ such that every set of the form $U_\nu(\theta)$ can be covered by at most $C(\nu)$ sets of diameter a_ν or less. Then, for every $k \geq 1$ one has

$$P_\theta(\theta \notin B_k) \leq 2 \sum_{\nu=1}^{k} \beta(\nu)C(\nu).$$

Proof. Fix a particular θ. For each ν, θ belongs to one of the sets $A_{\nu,i}$, $i \in I_\nu$. The set in question will be denoted $A_{\nu,0}$. Let J^*_ν be the set of indices $j \in I_\nu$ such that $A_{\nu,j}$ intersects $V_\nu(\theta) = \{t \in \Theta; W(\theta,t) < b_{\nu-1}\}$. If $i \in J^*_\nu$, let

$J_\nu^*(i)$ be the set of indices $j \in J_\nu^*$ for which (i,j) is a ν-distant pair. Let ψ_ν^* be the infimum

$\psi_\nu^* = \inf_j \{\varphi_{\nu,0,j}; j \in J_\nu^*(0)\}$. When $\theta \in B_{\nu-1}$ one has $B_{\nu-1} \subset V_\nu(\theta)$ and therefore $\psi_\nu^* \leq \psi_{\nu,0}$. Thus, if χ_ν is the indicator of the set of x such that $\theta \in B_\nu$ one may write $\chi_{\nu-1}(1-\chi_\nu) \leq \chi_{\nu-1}(1-\psi_\nu^*)$. This gives

$$1 - \chi_k = (1 - \chi_1) + \chi_1(1 - \chi_2) + \ldots + \chi_{k-1}(1 - \chi_k)$$
$$\leq \sum_{\nu=1}^k (1-\psi_\nu^*).$$

According to the selection of the tests $\varphi_{\nu,i,j}$, the expectation $E_\theta(1-\psi_\nu^*)$ does not exceed $2\beta(\nu)N_\nu$ for an N_ν which is the number of sets $A_{\nu,i}$ which intersect $V_\nu(\theta)$. Suppose that $N_\nu \geq C(\nu) + 1$. Then one can cover the set $U_\nu(\theta)$ by a certain number $N'_\nu < N_\nu$ of sets $A'_{\nu,i}$ such that diam $A'_{\nu,i} \leq a_\nu$. These together with the $A_{\nu,j}$ which do not intersect $V_\nu(\theta)$, form a cover whose cardinality is strictly inferior to that of the partition $\{A_{\nu,i}; i \in I_\nu\}$. This cannot be since I_ν had the minimum possible cardinality. Thus $N_\nu \leq C(\nu)$ and the proof of the proposition is complete.

To construct estimates $\hat{\theta}$ of θ one can proceed by selecting any integer m and choosing a point $\hat{\theta}$ in the last of the B_j, $j \leq m$ which is nonempty. This gives the following corollary.

Corollary. Let g be any monotone increasing function from $[0,\infty)$ to $[0,\infty)$. Define an estimate

$\hat{\theta}$ as indicated. Then

$$E_\theta g[W(\hat{\theta},\theta)] \leq g(b_m) + 2 \sum_{0 \leq \nu \leq m-1} g(b_\nu) C(\nu+1) \beta(\nu+1).$$

This follows from Proposition 1 by simple algebra.

The occurrence of the numbers $C(\nu)$ suggests that the preceding Proposition 1 will be usable when Θ metrized by W is of finite dimension in a sense which is very close to the one introduced by Kolmogorov in [7].

The definition used in the sequel is as follows.

Definition 1. Let $d_0 \geq 0$ be a prescribed number. The set Θ has (W, d_0) dimension at most D if every subset $S \subset \Theta$ whose W diameter does not exceed $2d \geq 2d_0$ can be covered with no more than $C = 2^D$ sets of diameter at most d.

The number D is related to the ordinary dimension. For instance, in R^k with the maximum coordinate norm, cubes of diameter $2d$ can be covered with 2^k cubes of diameter d. The introduction of the lower bound d_0 is useful in several arguments where the high dimensionality of a set Θ would show up only for covers by very small sets. If in the present instance, $0 \leq d_0 < a_\nu < b_\nu$ then the number $C(\nu)$ of sets needed to cover a ball $U_\nu(\theta)$ is at most 2^{kD} for the smallest integer k such that $2^k a_\nu \geq 2[b_\nu + a_\nu]$.

3. Using a preliminary estimate. Consider a set Θ, with a metric W, and two experiments $\mathcal{E} = \{P_\theta; \theta \in \Theta\}$ and $\mathcal{J} = \{Q_\theta; \theta \in \Theta\}$, both indexed by

Θ. The P_θ are measures on a σ-field $\mathcal{A}$. Similarly the Q_θ are measures on a σ-field $\mathcal{B}$. The two σ-fields $\mathcal{A}$ and $\mathcal{B}$ are not assumed to be related in any manner. For any set $S \subset \Theta$ let $\mathcal{E}_S = \{P_\theta; \theta \in S\}$ and $\mathcal{J}_S = \{Q_\theta; \theta \in S\}$. Using the notations and definitions of [3], intuition suggests the following. Suppose that $\mathcal{E}$ provides estimates $\hat{\theta}$ of θ which are such that $P_\theta\{W(\hat{\theta}, \theta) \geq b_0\}$ is small. Suppose also that if S is a set of diameter b_1 substantially larger than b_0, the deficiency $\delta(\mathcal{E}_S, \mathcal{J}_S)$ is small. Then $\delta(\mathcal{E}, \mathcal{J})$ ought to be small.

The purpose of the present section is to prove an assertion of this nature valid when Θ is subjected to a dimensionality restriction. We do not claim that the result given here is the best of its type and even hope to be able to improve it following the lines suggested by Bayesian approach. However, at this time, other arguments still need a substantial amount of "debugging."

A construction analogous to the one given here was used for the proof of Theorem 2, Section 11 of [3]. However, in that theorem, we use "insufficiency" instead of "deficiency." This allows many simplifications.

If Θ is a set metrized by W, call partition of unity on (Θ,W) any family $\{u_j, j \in J\}$ of continuous numerical functions such that $0 \leq u_j \leq 1$ and $\Sigma_j u_j = 1$. The following lemma will be a technical tool for other arguments.

Lemma 1. Let Θ be a finite set metrized by the distance W. Let a and b be two numbers such that $0 \leq a < b$. Assume that there is some number C such that every set of diameter $4b + 2a$ can be covered by no more than C sets of diameter inferior or equal to b. Then there are partitions of unity $\{u_j; j \in J\}$ such that

(i) each u_j satisfies the Lipschitz condition
$$|u_j(s) - u_j(t)| \leq \frac{1}{b} W(s,t).$$

(ii) For each set $S \subset \Theta$ of diameter diam $S \leq 2a$, the number of indices j such that $u_j(s) > 0$ for some $s \in S$ does not exceed C.

(iii) The support of each u_j has diameter at most $3b$.

Proof. Let $\{A_j; j \in J\}$ be a partition of Θ by sets A_j satisfying the requirement diam $A_j \leq b$. Assume that J has the minimum possible cardinality under the circumstances. For each A_j let B_j be the set of points of Θ whose distance to A_j is at least b. If $s \in \Theta$, let $W(s,A) = \inf\{W(s,t); t \in A\}$. Consider the function v_j defined by

$$v_j(s) = \frac{W(s,B_j)}{W(s,A_j) + W(s,B_j)}.$$

This is unity on A_j and zero on B_j. A simple computation shows that, for any pair (s,t) one has

$$b|v_j(s) - v_j(t)| \leq W(s,t).$$

Since Θ is finite, so is J. Thus we can pretend that J is a set $J = \{0,1,2,\ldots,n\}$ of integers.

Let $u_0 = v_0$. If u_j has been defined for $j = 0,1,2,\ldots,k$, define u_{k+1} by

$$u_0 + u_1 + \ldots + u_k + u_{k+1} = (u_0 + u_1 + \ldots + u_k + v_{k+1}) \wedge 1.$$

The functions u_j so defined satisfy the same Lipschitz condition as the v_j. Also $\Sigma_j u_j = 1$ since $\Sigma v_j \geq 1$. Clearly $u_j \leq v_j$. Hence u_j has a support of diameter at most $3b$.

Finally, let S be a set of diameter $2a$. If $u_j(s) > 0$ for some $s \in S$, then A_j is contained in the set S' of points whose distance to S is at most $2b$. Since S' has diameter at most $4b + 2a$, it could be covered by C sets of diameter b. Thus, as in the argument of Proposition 1, the number of indices j for which $\sup_S\{u_j(s); s \in S\} > 0$ cannot exceed C.

This lemma will now be used to provide an interpolation formula which pieces together transitions obtained on small subsets of Θ. In order to state an appropriate theorem we need to specify what is meant by an estimate available on $\mathcal{E}$. For this purpose, suppose that Θ is metrized by W. Let Γ be the space of bounded continuous functions on (Θ, W). An estimate available on $\mathcal{E}$ is then a transition from $L(\mathcal{E})$ to the dual Γ^* of Γ according to the definitions and conventions of [3]. Even if the estimate is written $\hat{\theta}$ a symbol such as $P_\theta\{W(\hat{\theta}, \theta) > a\}$ is to be understood as a risk according to the definition used in [3].

Theorem 1. _Let_ $\mathcal{E} = \{P_\theta; \theta \in \Theta\}$ _and_ $\mathcal{F} = \{Q_\theta; \theta \in \Theta\}$ _be two experiments indexed by the set_ Θ. _Assume that_ Θ _is metrized by W, that_ $0 \leq a < b$ _are_

given and that any subset of diameter 4b + 2a of
Θ can be covered by no more than C sets of diameter
b.

Furthermore, assume that
(i)   if S ⊂ Θ has a diameter 3b then the
      deficiency $\delta(\mathcal{E}_S, \mathcal{J}_S)$ does not exceed ϵ_1,
(ii) there is an estimate $\hat{\theta}$ available on $\mathcal{E}$ such
      that $P_\theta\{W(\hat{\theta},\theta) > a\} \leq \epsilon_2$ for all $\theta \in \Theta$.
Then

$$\delta(\mathcal{E},\mathcal{J}) \leq \epsilon_1 + \epsilon_2 + \frac{1}{2}(\frac{a}{b})C.$$

Proof. Since the deficiency in question is the
supremum of deficiencies computed on finite subsets,
one may as well assume that Θ is finite. This will
be done below. In this case an estimate available
on $\mathcal{E}$ is representable by a Markov kernel $K(x,B)$
defined for $x \in \chi$ and for all subsets B of Θ. One
can inflate the experiment $\mathcal{E}$ taking for new measures
the semidirect products $\hat{P}_\theta(A \times B) = \int_A K(x,B) P_\theta(dx)$.
The inflated experiment $\hat{\mathcal{E}}$ is equivalent to $\mathcal{E}$. On
it, the estimate is nonrandomized. Thus, it is
enough to look at estimates $\hat{\theta}$ which are measurable
functions with values in Θ.

Let $\{A_j; j \in J\}$ be a partition of Θ by sets
such that diam $A_j \leq b$. Assume that the partition
in question has the smallest possible cardinality
and let $\{u_j; j \in J\}$ be the Lipschitzian partition
of unity provided by Lemma 1. For each index
$j \in J$, let T_j be a transition from $L(\mathcal{E})$ to $L(\mathcal{J})$
such that if $\overline{A}_j$ is the set of points at distance at
most b of A_j then

$$\frac{1}{2} \sup_{\theta \in \bar{A}_j} \| Q_\theta - T_j P_\theta \| \leq \epsilon_1 .$$

Since $\bar{A}_j$ has diameter at most 3b such a transition exists by assumption.

Consider the transition $T = \Sigma_j T_j f_j$ where f_j is the function $f_j(x) = u_j[\hat{\theta}(x)]$ and where an image $T_j f_j \bullet \mu$ is defined as the image for T_j of the measure $f_j \bullet \mu$ which has density f_j with respect to μ.

Fix a particular $\theta \in \Theta$. Let S be the set $S = \{s: W(\theta,s) \leq a\}$ and let J_θ be the set of indices $j \in J$ such that $u_j(s) > 0$ for some $s \in S$. Let P'_θ be the truncation of P_θ to the set $\{\hat{\theta} \in S\}$. The image $Q^*_\theta = \Sigma T_j f_j \bullet P_\theta$ is a probability measure which may be written $Q^*_\theta = Q' + Q'' + Q'''$ with

$$Q' = \Sigma \{T_j\, f_j \bullet P'_\theta\, ; \, j \in J_\theta\}$$

$$Q'' = \Sigma \{T_j\, f_j \bullet (P_\theta - P'_\theta)\, ; \, j \in J_\theta\}$$

$$Q''' = \Sigma \{T_j\, f_j \bullet P_\theta\, ; \, j \notin J_\theta\}.$$

The total mass of Q''' is the sum of the integrals $f_j P_\theta$ for $j \notin J_\theta$. These are also equal to $f_j(P_\theta - P'_\theta)$.

Since $\| P_\theta - P'_\theta \| \leq \epsilon_2$ by assumption, one can assert that $\| Q'' \| + \| Q''' \| \leq \epsilon_2$. Now consider Q' and let α_j be such that $f_j P'_\theta = \alpha_j \| P'_\theta \|$. The Lipschitz condition of Lemma 1 implies that $|f_j - \alpha_j| \leq a/b$. Thus we may write

$$Q' = \Sigma T_j f_j \cdot P'_\theta = \Sigma \alpha_j T_j P'_\theta + \sum_j T_j(f_j - \alpha_j) \cdot P'_\theta,$$

all sums being taken for $j \in J_\theta$. Since the cardinality of J_θ is at most C, the last term on the right has a norm at most $C(a/b)$. For the first term on the right, note that $\|T_j P_\theta - Q_\theta\| \leq 2\epsilon_1$. Therefore

$$\|Q_\theta - \Sigma \alpha_j T_j P'_\theta\| \leq 2\epsilon_1 + \epsilon_2.$$

Assembling terms gives the result as stated.

Remark 1. The number C which appears in the preceding statement is related to the dimension of Θ. If this is defined as in Section 2, one can assert that $C \leq 2^{4D}$, the coefficient D being the dimension for (W,b). The argument given in [3], Theorem 2, Section 11 also relies on a dimensionality restriction. However, in that case it is not necessary to introduce the Lipschitz smoothing.

Remark 2. In the statement of Proposition 1 and its corollary we have assumed finiteness of the set Θ. Trivial modifications of the argument make it perfectly valid under Assumption 1 above whenever the family P_θ is dominated. However, one can also note that the reduction given at the beginning of the proof of Theorem 1 could be used to extend the validity of Proposition 1 to arbitrary sets Θ (subject to the dimension in Assumption 1). It is enough for this purpose to use for the definitions of tests or estimates the general definition by transitions used in Theorem 1. Since part of the purpose of Proposition 1 is to allow application of Theorem 1 this is quite legitimate.

4. <u>Existence of estimates for independent observations</u>. In this section we consider a particular experiment $\mathcal{E} = \{P_\theta; \theta \in \Theta\}$ given by measures P_θ on a measurable space $(\mathcal{X}, \mathcal{A})$. It will be assumed that there is a certain set of indices L and that $\{\mathcal{X}, \mathcal{A}, P_\theta\}$ is the direct product of components $\{\mathcal{X}_\iota, \mathcal{A}_\iota, P_{\theta,\iota}\}$ for $\iota \in L$.

In this product situation it turns out that it is convenient to use a particular distance H derived from Hellinger distances as follows. For $\iota \in L$ let h_ι be the Hellinger distance defined by

$$h_\iota^2(s,t) = \tfrac{1}{2}\int(\sqrt{dp_{s,\iota}} - \sqrt{dp_{t,\iota}})^2.$$

Let $H^2(s,t) = \sum_\iota h_\iota^2(s,t)$ and let

$$r^2(s,t) = \tfrac{1}{2}\int(\sqrt{dP_s} - \sqrt{dP_t})^2 = 1 - \prod_\iota[1-h_\iota^2(s,t)].$$

It would be natural to use the distance r, or even more natural to use the related L_1-norm $\tfrac{1}{2}\|P_s - P_t\|$. However, one has always $1-r^2 \leq \exp\{-H^2\}$ and, if $h_\iota^2 \leq \alpha < 1$ for all $\iota \in L$, the opposite inequality $r^2 \leq 1 - \exp\{-\beta H^2\}$ with $\alpha\beta = -\log(1-\alpha)$. Thus, at least when all the h_s are small the distance H is not entirely unreasonable. If on the contrary, the individual h_ι may be large, H is unsatisfactory. It may happen, for instance that $H(s,t) \leq 2$ but that P_s and P_t are disjoint. In such a case it would be preferable to use loss functions such as $W(s,t) = -\log\|P_s \wedge P_t\|$. Unfortunately the argument given below does not apply to such a case. Thus the result of this section should be considered as intended for situations where the components of the experiment $\mathcal{E}$ provide individually very little information.

Given the set Θ and the pseudo metric H one can always identify s and t if $H(s,t) = 0$. Also, if $H(s,t) = \infty$, then P_s and P_t are disjoint.

A pairwise testing procedure analogous to that of Proposition 1 will provide estimates $\hat{\theta}$ such that $P_\theta\{H(\hat{\theta},\theta) < \infty\} = 1$ for all θ.

Thus we need to consider only sets Θ on which H remains finite and is actually a metric. Finally, arguing as in the proof of Theorem 1, one may assume that Θ is a finite set.

In summary, we shall proceed under the following restriction.

Assumption 3. The set Θ is a finite set metrized by H. There is a number $D \geq 1$ such that any subset of Θ having a diameter $d > 2^{-6}$ can be covered by no more than 2^D sets of diameter $(d/2)$.

Under this assumption, consider sequences (a_ν, b_ν), $\nu = 0,1,2,\ldots,m$, and associated minimal partitions $\{A_{\nu,i}; i \in I_\nu\}$ as in Section 2.

Select two numbers, $a \in [0,1]$ and $c \geq 1$ in such a way that $w = e^{-c} + 2a(2 - a^2)^{\frac{1}{2}} \leq 1$.

Lemma 2. For any sequence $b_0 > b_1 > \ldots > b_m$ and any $a \in [0,1]$ such that $a \leq b_m$ there are partitions $\{A_{\nu,i}; i \in I_\nu\}$ such that for all ν-distant pairs (i,j) the sum of error probabilities $\pi(A_{\nu,i}; A_{\nu,j})$ satisfies the inequality

$$\pi(A_{\nu,i}, A_{\nu,j}) \leq [w(2-w)]^{n/2}$$

for an integer n at least equal to the integer part of $b_\nu^2/(c+1)$.

Proof. Let $\{A_{\nu,i}; i \in I_\nu\}$ be a minimal partition by sets of diameter at most $a_\nu \geq a$. For a ν-distant pair (i,j) take $s \in A_{\nu,i}$ and $t \in A_{\nu,j}$ so that $H^2(s,t) \geq b_\nu^2$. Write $h_\ell = h_\ell(s,t)$ for simplicity. One can partition the set of indices L in $(n+1)$ disjoint subsets L_k, $k = 1, 2, \ldots, n+1$ in such a way that if $1 \leq k \leq n$ then $S_k^2 = \Sigma\{h_\ell^2; \ell \in L_k\} \geq c$.

Let $P_{\theta,k}$ be the product $P_{\theta,k} = \prod_{\ell \in L_k} P_{\theta,\ell}$.

Consider $P_{s,k}$ and $P_{t,k}$ for $k \leq n$. By construction the affinity between the two is at most e^{-c}. Therefore, $\|P_{s,k} \wedge P_{t,k}\| \leq e^{-c}$.

Let $a_{\nu,i,k}$ be the diameter of $A_{\nu,i}$ for the Hellinger distance $H_k(s,\theta) = h(P_{s,k}, P_{\theta,k})$. Suppose that both $a_{\nu,i,k}$ and $a_{\nu,j,k}$ do not exceed a. Note that for any two probability measures P and Q one has

$$\tfrac{1}{2}\|P - Q\| \leq h(P,Q)[2 - h^2(P,Q)]^{\tfrac{1}{2}}.$$

Thus, with the assumption $\max(a_{\nu,i,k}, a_{\nu,j,k}) \leq a$ one can assert that the observations with indices $\ell \in L_k$, $k \leq n$ provide tests with sum of error probabilities at most equal to $w = e^{-c} + 2a(2-a^2)^{\tfrac{1}{2}}$. The assertion $\pi(A_{\nu,i}, A_{\nu,j}) \leq [w(2-w)]^{n/2}$ for the whole product follows as usual. Thus, it will be enough to show that the inequalities $a_{\nu,i,k} \leq a$ can be achieved. However, this is certainly the case if the partitions $\{A_{\nu,i}; i \in I_\nu\}$ are independent of ν and taken for a sequence $a_\nu \equiv a$. This completes the proof of the lemma.

At this point it is convenient to introduce auxiliary notations.

Having selected a and c writing $w = e^{-c} + 2a(2 - a^2)^{\frac{1}{2}}$, let $e^{-2\gamma} = w(2 - w)$. Then the bound of Lemma 2 could be rewritten

$$\pi(A_{\nu,i}, A_{\nu,j}) \le e^{-\gamma n} \le e^\gamma \exp\left\{-\frac{\gamma b_\nu^2}{c+1}\right\}.$$

To specify a sequence $\{b_\nu\}$, choose a number $q > 1$ and let $b_{\nu-1} = qb_\nu$ with a b_0 at least equal to the diameter of Θ.

The operation described in Section 2 with sets $A_{\nu,i}$ of diameter at most $a \ge (1/125)$ and stopping at a certain integer m <u>will be called Procedure</u> $[c,a,q,m]$.

To evaluate the performance of this Procedure according to Proposition 1, it will be sufficient to obtain bounds on the covering numbers $C(\nu)$. By Assumption 3 any set of diameter $d > a$ can be covered by at most $(2\frac{d}{a})^D$ sets of diameter a. For a ball of radius $b_{\nu-1} + a$ the number of sets needed to cover is then at most

$$C(\nu) \le \left[r\frac{b_{\nu-1}}{a}\right]^D$$

provided that $r \ge 4 + (a/b_{\nu-1})$.

Since, typically, the integer m will be chosen such that b_{m-1} is substantially larger than a the coefficient r will in fact be very close to 4.

For simplicity in the following derivations we shall also use the notations

(i) $s = D/2$

(ii) $u = s(c+1)q^2\gamma^{-1}$

(iii) $z = b_m^2 u^{-1}$

(iv) $A_0 = \left(\frac{r}{a}\right)^{2s} e^\gamma s^s \left(\frac{c+1}{\gamma}\right)^s \frac{q^{2s}}{\log q}$

(v) $A = \frac{1}{s} \log A_0$
$= \log\left(\frac{c+1}{\gamma} \frac{r^2}{a^2}\right) + \log s$
$+ \frac{1}{s}\left[\gamma + \log\left(\frac{q^{2s}}{\log q}\right)\right]$.

(vi) z_0 is the solution $z_0 \geq 1$ of the equation $z_0 = A + \log z_0$

(vii) $f(z) = z + A_0 \int_z^\infty x^s \exp\{-sx\}dx$.

Proposition 2. <u>Let Assumption 3 be satisfied and let the numbers a, c and q be fixed. Assume a $2\sqrt{2} \leq 1$. Then there is an integer m and a number b_m such that the procedure [c,a,q,m] yields an estimate $\hat{\theta}$ for which</u>

$$E_\theta H^2(\hat{\theta},\theta) \leq u\, f(z_0)$$
$$\leq s\, \frac{c+1}{\gamma} q^2 z_0\left[1 + \frac{1}{s(z_0-1)}\right].$$

Proof. Applying Proposition 1 and its corollary to the bounds for $\pi(A_{\nu,i}, A_{\nu,j})$ and $C(\nu)$ described above one obtains the relation

$$E_\theta H^2(\hat{\theta},\theta) \leq b_m^2 + 2 e^\gamma T_m$$

with

$$T_m = \sum_{0 \le k \le m-1} b_k^2 \left(\frac{rb_k}{a}\right)^D \exp\left\{-\frac{\gamma}{c+1} b_{k+1}^2\right\}.$$

Introducing the term u defined above one sees that the sum T_m resembles the integral

$$I_m' = \frac{r^D}{a} \int_0^\infty (b_m q^2)^{2+D} \exp\left\{-\frac{D}{2u} b_m^2 q^{2\nu}\right\} d\nu.$$

Introducing the variable $z = b_m^2 u^{-1}$ and changing variables, the expression $b_m^2 + 2 e^\gamma I_m$ can be put in the form

$$K(z) = u\, f(z)$$

with $f(z) = z + A_0 \int_z^\infty x^S \exp\{-sx\}\, dx$.

To relate this to the sum T_m and to select an appropriate value of b_m^2 we need a lower bound on the coefficient A. A crude one is obtainable as follows. By construction $e^{-2\gamma} \ge e^{-c}(2-e^{-c})$. Thus $2\gamma \le c$. Also $2a\sqrt{2}$ cannot be much larger than unity.

Assuming $2a\sqrt{2} \le 1$, one can assert that $(c+1)\gamma^{-1} > 2$ and $a^{-2} \ge 8$. Substituting this in the expression of A one sees that, even for $D = 1$ the number A must be larger than 6.9. Thus $A_0 \exp\{-s\}$ is much larger than unity.

Consider then the first and second derivatives f' and f'' of the function f. The function f' is given by

$$f'(z) = 1 - A_0\, z^S \exp\{-sz\}.$$

The second derivative $f''(z)$ is proportional to

$$z^{s-1} \exp\{-sz\}(z-1).$$

It follows that f' achieves a minimum at the point $z = 1$. This minimum is equal to $f'(1) = 1 - A_0 \exp\{-s\} = 1 - \exp\{s(A-1)\} < 0$. Thus, in the interval $(0,\infty)$, the derivative f' vanishes at two points which straddle the point $z = 1$. One concludes that the function f achieves a minimum at that point $z_0 \geq 1$ which is solution of the equation $z_0 = A + \log z_0$.

To obtain a bound on $f(z_0)$, note that one may write

$$\begin{aligned} f(z_0) &= z_0 + \int_{z_0}^{\infty} \left(\frac{x}{z_0}\right)^s \exp\{-s(x - z_0)\}dx \\ &= z_0[1 + \int_0^{\infty} (1+\xi)^s \exp\{-sz_0\xi\}d\xi] \\ &\leq z_0[1 + \int_0^{\infty} \exp\{-s(z_0 - 1)\xi\}d\xi] \\ &= z_0\left\{1 + \frac{1}{s(z_0 - 1)}\right\}. \end{aligned}$$

It should be mentioned that the relations $z_0 \geq 1$, $z_0 = A + \log z_0$ imply $z_0 \geq A + \log A$. Therefore the denominator $s(z_0 - 1)$ in this expression cannot be smaller than 2.9.

To complete the proof it remains to show that the integral I_m bounds the sum T_m. For this purpose let $\varphi(\nu)$ be defined by

$$\varphi(\nu) = q^{2\nu(1+s)} \exp\{-sz_0 q^{2\nu}\},$$

in the interval $0 \leq \nu \leq m$. This is a decreasing

function of ν, in that interval, provided only that $z_0 \geq 1 + s^{-1}$. However, since $2s \geq 1$ and since $z_0 \geq 6.9$, the requirement in question is certainly satisfied.

It follows from this that

$$\sum_{\nu=1}^{m} \varphi(\nu) \leq \int_0^\infty \varphi(\nu) d\nu.$$

Substituting this in the formulas for I_m and T_m one sees that $I_m \geq T_m$ for the selected value z_0. The desired conclusion follows from this and the remark that the value $b_m^2 = u z_0$ is a possible value in our decreasing sequence $\{b_\nu^2\}$.

Proposition 3. <u>Let Assumption (3) be satisfied and let the numbers a, c, and q be fixed. Assume a $2\sqrt{2} \leq 1$ and let B_m be the confidence set provided by a procedure [c,a,q,m] with $b_m^2 = uz$, $z \geq 1$. Then</u>

$$P_\theta\{\theta \notin B_m\} \leq A_0 \int_z^\infty x^{s-1} \exp\{-sx\} dx.$$

<u>Proof</u>. Substituting in Proposition 1 the bounds for $C(\nu)$ and $\pi(A_{\nu,i}, A_{\nu,j})$ used above one sees that the probability in question is bounded by

$$2 e^\gamma \frac{r^{2s}}{a} u^s \sum_{\nu=1}^{m} (z q^{2\nu})^s \exp[-s z q^{2\nu}].$$

The result follows upon replacement of the discrete sum by an integral. This replacement can be justified exactly as in **the** proof of Proposition 2. Hence the result.

Propositions 2 gives a bound on the risk

$E_\theta H^2(\hat{\theta},\theta)$. However, this bound is expressed in terms of quantities u and z_0 which involve the numbers a, c, q and D in a complicated manner. The following result gives a bound involving only universal constants and the dimension D.

Proposition 4. <u>Let Assumption (3) be satisfied. There is a procedure [c,a,q,m] for which</u>

$$E_\theta H^2(\hat{\theta},\theta) \le (4.02)D \log D + (53.5)D + (23.1),$$

<u>for all</u> $\theta \in \Theta$.

Proof. Consider the bound

$$K(z_0) = s\left(\frac{c+1}{\gamma}\right)q^2 z_0\left[1 + \frac{1}{s(z_0-1)}\right]$$

of Proposition 2. The number $z_0 \ge 1$ is the root of the equation $z_0 = A + \log z_0$. It is an increasing function of A. Also, the bound $K(z_0)$ is an increasing function of z_0 provided that $z_0 \ge 1 + s^{-\frac{1}{2}}$. Since $z_0 \ge A \ge 6.9$, this last requirement is certainly satisfied.

To bound A, let us recall that $\gamma = -\frac{1}{2} \log w(2-w)$ with $w = e^{-c} + 2a(2-a^2)^{\frac{1}{2}}$. For small a the term w differs little from the slightly larger $w_1 = e^{-c} + a 2\sqrt{2}$. We shall replace w by w_1. To obtain a numerical expression we need to choose the number c.

We do not know what the optimal value is and will, arbitrarily, put c = 3.69. Also we shall, arbitrarily, fix a so that $e^{-c} = a\ 2\sqrt{2}$. Then, $w_1 = 2\ e^{-c}$ and a^2 is approximately equal to $78\ 10^{-6}$. In particular $125a \ge 1$ so that Assumption (3) is applicable.

Simple arithmetic shows that, for this choice of constants the quantity A is certainly larger than 12.9. Also the ratio (a/b_m) is less than $4 \cdot 10^{-4}$. Thus $r \leq 4.002$ and finally

$$\log\left[\left(\frac{c+1}{\gamma}\right)\frac{r^2}{a^2}\right] \leq 13.627.$$

To finish the specification of the procedure we need to select the value of q. Here again we do not know the optimal value. The formula for the bound $K(z_0)$ seems to indicate that q should be a decreasing function of the dimension D and that it should be close to unity.

We shall take, arbitrarily, $q = 1.2$. With this choice the procedure is entirely specified. Simple algebra shows that

$$12.9 \leq A \leq 14 + \log \frac{D}{2} + \frac{5.736}{D},$$

and that, for $A \geq 12.9$ the root z_0 satisfies the inequality $z_0 \leq (1.22)A$. Upon substitution of these numbers in $K(z_0)$ one obtains the result as stated thereby completing the proof of the proposition.

5. <u>Identically distributed observations.</u> In this section we shall keep all the assumptions in vigor in Section 4, but assume in addition that the spaces $(\mathcal{X}_\iota, \mathcal{Q}_\iota, P_{\theta,\iota})$, $\iota \in L$ are all copies of each other. In other words, the observations are independent and identically distributed.

The result which corresponds to Proposition 4 in this case can be stated as follows.

<u>Proposition 5.</u> <u>Let Assumption (3) be satisfied.</u>

Let the spaces $(\mathfrak{X}_\ell, \mathcal{Q}_\ell, P_{\theta,\ell})$ be copies of one space $(\mathfrak{X}_0, \mathcal{Q}_0, P_{\theta,0})$, then there are estimates $\tilde{\theta}$ such that

$$E_\theta H^2(\hat{\theta}, \theta) \le 19 + 37D.$$

Proof. The proof follows very closely the outline of the proof of Proposition 4. However, in the present case one may easily select partitions $\{A_{\nu,i}; i \in I_\nu\}$ with diameters diam $A_{\nu,i} \le a_\nu$ depending on ν and still obtain the result of Lemma 2. Indeed, the partitioning $L = \bigcup L_k$; $k = 1, 2, \ldots, n, n+1$ of L can be selected so that all the L_k, $k \le n$ have the same cardinality. The proof of Lemma 2 depends only on the fact that for the product measures $P_{\theta,k} = \Pi_{\ell \in L_k} P_{\theta,\ell}$ the Hellinger distance diameter of the $A_{\nu,i}$ is not more than a. Now let h^2 be the square diameter of $A_{\nu,i}$ for the square distance

$$H_k^2(s,t) = \Sigma\{h_\ell^2(s,t); \ell \in L_k\}.$$

Since there are $n+1$ pieces L_k one can write $a_\nu^2 = nh^2 + \nu^2$, with $0 \le \nu^2 < c$. The inequality $h^2 \le a^2$ is certainly satisfied if $a_\nu^2 - \nu^2 \le na^2$. Since n is at least equal to the integer part of $b_\nu^2/(c+1)$, a simple computation shows that it is possible to take any value a_ν such that

$$\frac{b_\nu^2}{a_\nu^2} \ge \frac{1}{a^2}\left[1 + \frac{(c+1)a^2}{a_\nu^2}\right].$$

Proceeding as in Section 4, with $b_{\nu-1} = qb_\nu$, one sees that in the present case, the number of sets

$A_{\nu,i}$ needed to cover the ball of radius $a_\nu + b_{\nu-1}$ can be bounded independently of ν. This removes the terms of type x^s in the Gamma integrals of Section 4. A computation based on the choice $c = 3.69$ and $q = 1.2$ of Section 4 gives the bound

$$E_\theta H^2(\hat{\theta},\theta) \leq (18.5) + (36.8)D,$$

hence the result as stated.

In a previous paper, this author had already stated a result analogous to the one described by Proposition 5. Unfortunately, the argument given there is incorrect. We take covers $\{A_{\nu,i}\}$ which appear to be selected after $B_{\nu-1}$ is obtained and we count only the sets necessary to cover $B_{\nu-1}$ instead of the possibly larger balls of radius $a_\nu + b_{\nu-1}$. In spite of the necessary corrections, the present result compares favorably with the previous assertion because of a better choice of constants c and q.

The bound which appears in [4] is $16 \log 6c$ for a constant c which denotes the number of sets of diameter ε needed to cover a set of diameter $\varepsilon 2^{11/2}$. Thus the $(\log c)$ in question is to be compared with the present $(5.5)D$.

6. <u>Approximations by Poisson processes.</u> Let ℬ denote a ring of subsets of a set 𝒴. Let μ be a positive finite measure defined on ℬ. By Poisson process of base μ we shall understand a family $\{X_B; B \in ℬ\}$ of random variables such that

(i) If $B_1, B_2, \ldots, B_k$ is a finite family of elements of ℬ formed of pairwise disjoint sets then the corresponding $\{X_{B_i}\}$ are independent.

(ii) For each $B \in \mathcal{B}$ the variable X_B has a Poisson distribution with expectation $EX_B = \mu(B)$.

Let Θ be a set. For each $\theta \in \Theta$ let μ_θ be a positive measure on $\mathcal{B}$ and let Q_θ be the distribution of the Poisson process of base μ_θ. The family $\{Q_\theta; \theta \in \Theta\}$ is an infinitely divisible experiment which we shall call the Poisson experiment of base $\{\mu_\theta; \theta \in \Theta\}$.

Poisson experiments arise on their own merits. They also arise as possible approximations for direct products of the type studied in Section 4. The paper [4] contains some indications on the possibility of such approximations for independent identically distributed variables. This possibility, which is already useful in the isonomous case, is much more appealing when the observations are not identically distributed.

Suppose then that $\mathcal{E} = \{P_\theta; \theta \in \Theta\}$ is a direct product of experiments $\mathcal{E}_\ell = \{p_{\theta,\ell}; \theta \in \Theta\}$, with $\ell \in L$. For simplicity assume that L is finite. The measures $p_{\theta,\ell}$ are carried by spaces $(\mathcal{X}_\ell, \mathcal{G}_\ell)$, however, it is always possible to rearrange them <u>on a common space</u> $(\mathcal{Y}, \mathcal{B})$ <u>in such a way that the likelihood ratios of the type</u> $(dp_{t,\ell}/dp_{s,\ell})$ <u>are independent of</u> ℓ. This can be done by passage to the space of likelihood ratio processes, or by taking for $(\mathcal{Y}, \mathcal{B})$ the direct sum of the $(\mathcal{X}_\ell, \mathcal{G}_\ell)$.

Assuming that the experiments $\mathcal{E}_\ell$ are so reorganized, let $\mu_\theta = \Sigma_\ell p_{\theta,\ell}$ and let $\pi = \{Q_\theta; \theta \in \Theta\}$

be the Poisson experiment based on the family $\{\mu_\theta; \theta \in \Theta\}$. Define metrics h_ℓ by $h_\ell^2(s,t) = \frac{1}{2}\int (\sqrt{dp_{s,\ell}} - \sqrt{dp_{t,\ell}})^2$ and let

$$H^2(s,t) = \sum_\ell \int h_\ell^2(s,t) = \frac{1}{2}\int(\sqrt{d\mu_s} - \sqrt{d\mu_t})^2.$$

Finally let

$$\beta(s,t) = \left[\sum_\ell h_\ell^4(s,t)\right] \exp\{-H^2(s,t)\},$$

and $\beta = \sup\{\beta(s,t); (s,t) \in \Theta \times \Theta\}$.

We shall assume that the space Θ satisfies the following dimensionality restriction.

Assumption 4. For every $d > 0$, subsets of Θ whose H-diameter does not exceed $2d$ can be covered by no more than 2^D sets of diameter inferior or equal to d.

In other words, the space Θ has $(H,0)$ dimension at most D.

With these definitions and assumptions one can state the following theorem, whose main failing is that one cannot escape feeling that much better results should be provable.

Theorem 2. For each $D \geq 1$ there is a function $\epsilon \to \gamma(D,\epsilon)$ such that $\gamma(D,\epsilon) \to 0$ as $\epsilon \to 0$ and for which the following relations hold.

Suppose that $\mathcal{E}$ is a direct product experiment with corresponding Poisson experiment $\mathcal{J}$. If Assumption 4 is satisfied then $\Delta(\mathcal{E},\mathcal{J}) \leq \gamma(D,\beta)$.

Proof. According to the usual arguments it is enough to prove the result for experiments indexed by sets Θ which are finite.

Let us show first that for any $\varepsilon \in (0, \frac{1}{2})$ one can select a β such that $\delta(\mathcal{E}, \mathcal{J}) \leq 7\varepsilon$. For this purpose note that Proposition 3 provides the possibility of finding a number a and estimates $\hat{\theta}$ such that $P_\theta\{H(\hat{\theta}, \theta) > a\} < \varepsilon$. This done, select $b > a$ so that $(\frac{a}{b})2^{6D} < \varepsilon$. According to Theorem 1 one will have $\delta(\mathcal{E}, \mathcal{J}) \leq 2\varepsilon + \varepsilon_1$ where ε_1 is the maximum of deficiencies $\delta(\mathcal{E}_S, \mathcal{J}_S)$ for sets S of diameter at most 3b. Using Assumption 4 again one sees that a set of diameter 3b can be covered by at most $K \leq \exp\{(\frac{3b}{\varepsilon} + 1)D \log 2\}$ sets of diameter ε. Suppose that $\{A_j; j=1,2,\ldots,K\}$ is such a cover of S. One can assume that it is a partition and take elements $\theta_j \in A_j$. Let then $\mathcal{E}'_S$ be the experiment $\mathcal{E}'_S = \{P'_\theta; \theta \in S\}$ with $P'_\theta = P_{\theta j}$ if $\theta \in A_j$. Operate similarly for $\mathcal{J}_S$ obtaining $\mathcal{J}'_S$. Since A_j has diameter ε one can write $\frac{1}{2}\|P'_\theta - P_\theta\| \leq 2\varepsilon$. Therefore, $\Delta(\mathcal{E}'_S, \mathcal{E}_S) \leq 2\varepsilon$ and similarly for $\mathcal{J}'_S$ and $\mathcal{J}_S$. Now $\mathcal{E}'_S$ and $\mathcal{J}'_S$ behave exactly as if their sets of indices had a cardinality at most K. Thus, according to Lemma 5, Section 8 of [3], there is a β so small that $\Delta(\mathcal{E}'_S, \mathcal{J}'_S) \leq \varepsilon$.

Putting all terms together we obtain that $\delta(\mathcal{E}, \mathcal{J}) \leq 7\varepsilon$. The reverse inequality, $\delta(\mathcal{J}, \mathcal{E}) \leq 7\varepsilon$, can be obtained by the same procedure. It is sufficient to note that since $\mathcal{J}$ is infinitely divisible Proposition 3 is still applicable to it. This concludes the proof of the theorem.

Approximations by Poisson processes often provide an easy method of study of limits of experiments. Examples of such approximations can be found for instance in [8].

However, one should note that the general structure of the proof of Theorem 2 does not rely very strongly on the fact that the approximating experiment is given by a Poisson process. Similar techniques could be used in other cases as well.

7. <u>Asymptotic normality</u>. Let $\mathcal{G} = \{G_t; t \in T\}$ be an experiment indexed by an arbitrary set T. It will be called <u>homogeneous</u> if the measures G_t, $t \in T$ are mutually absolutely continuous. For such an experiment one can take a particular $t_0 \in T$ and consider the logarithms

$$\Lambda(s) = \log \frac{dG_s}{dG_{t_0}}$$

as a stochastic process Λ for the distribution induced by the measure G_{t_0}. We shall call Gaussian experiment any homogeneous experiment $\mathcal{G}$ in which the process Λ is a Gaussian process.

If one takes the same measurable function $\Lambda(s)$ but induces the distribution of Λ by the measure G_t instead of G_{t_0}, the process obtained is also Gaussian, with the same covariance structure, but with expectations shifted by the amount $K(s,t)$ equal to the covariance for G_{t_0} of the pair $[\Lambda(s), \Lambda(t)]$.

Thus, to be precise one should say that $\mathcal{G}$ is homoschedastic Gaussian. Even though piecing together (by Theorem 1), local approximations by Gaussian experiments automatically lead to heteroschedastic Gaussian experiments we shall keep the simpler terminology.

The natural habitat of Gaussian experiments can

be described as follows. Let V be a real vector space with a Hilbertian norm, denoted $\|v\|$. On such a space there is a standard Gaussian linear process $v \to \langle Z, v \rangle$ such that $E\langle Z, v \rangle = 0$ and $E|\langle Z, v \rangle|^2 = \|v\|^2$. Take for G_0 the distribution of that process and take for G_v the measure whose density with respect to G_0 is $\exp\{\langle Z, \mathbf{v} \rangle - \frac{1}{2}\|v\|^2\}$.

If $\mathcal{G}$ is a Gaussian experiment indexed by T, according to our previous definition, one can put it in this standard form as follows. Let $t \to Z_0(t)$ be any real Gaussian process indexed by T and such that $E Z_0(t) = 0$. Form the process $Z(t) = Z_0(t) - Z_0(t_0)$. It has a certain covariance kernel K. Let V_0 be the space of signed measures with finite support on T. Define $\langle Z, v \rangle = \int Z(t) v(dt)$, and $\|v\|^2 = E|\langle Z, v \rangle|^2$ and complete V_0 for this norm. Then T can be identified with a subset of the completion V of V_0 and t_0 becomes the origin of V.

It should be clear that for a Hilbert space $(V, \|\circ\|)$, the corresponding Gaussian experiment is not uniquely defined since one has a wide choice of underlying probability spaces. However, the equivalence class of the experiment (in the sense of [3]) is well defined. Thus, by abuse of language, we shall speak of the Gaussian experiment indexed by the subset T of the Hilbert space $(V, \|\circ\|)$. The case where the function $v \to \|v\|$ is only a semi-norm should also be considered. It reduces to the preceding one by identification of two values v_1, v_2 if they give the same distributions to the observations.

In order to simplify studies of convergences

we shall use the following lemma which admits of various modifications obtainable by restricting the range of v to suitable cones of V.

Lemma 3. _Let V be a real vector space with a Hilbertian norm denoted_ $\|v\|$. _Let_ $\mathcal{G} = \{G_v;\ v \in V\}$ _be a homogeneous experiment indexed by V. Define a process L by_

$$L(v) = \log \frac{dG_v}{dG_0} + \tfrac{1}{2}\|v\|^2,$$

the distributions being induced by G_0. _Then_ $\mathcal{G}$ _is the standard Gaussian experiment on_ $(V, \|\circ\|)$ _if and only if_ $L(s+t) = L(s) + L(t)$ _almost surely for each pair_ (s,t) _of elements of V._

Proof. The condition is obviously necessary. Conversely, let $\{s_j;\ j = 1,2,\ldots,k\}$ be a fixed k-tuple of elements of V. The additivity condition implies that the relation $L(\Sigma\ \alpha_j s_j) = \Sigma\ \alpha_j\ L(s_j)$ holds almost surely for all k-tuples $\{\alpha_j\}$ which have rational components.

Letting $\Lambda(v) = \log(dG_v/dG_0)$, the homogeneity assumption implies $E \exp\{\Lambda(\Sigma\ \alpha_j s_j)\} = 1$ or equivalently

$$E \exp\{\Sigma\ \alpha_j\ L(s_j)\} = \exp\{\tfrac{1}{2}\|\Sigma\ \alpha_j s_j\|^2\}$$

for all rational k-tuples $\{\alpha_j\}$. If follows that the same relation holds also for arbitrary k-tuples of real numbers. As a result the process $v \to L(v)$ is a linear process. The distribution of $L(v)$ is Gaussian with expectation zero and variance $E|L(v)|^2 = \|v\|^2$. This concludes the proof of the desired result.

To describe asymptotic results consider a sequence $\{\mathcal{E}_n\}$ of experiments $\mathcal{E}_n = \{P_{s,n}; s \in S_n\}$ indexed by sets S_n. Let the Hilbert space $(V, \|\circ\|)$ be fixed. For each n let A_n be a map from S_n into V.

Definition 2. The sequence of maps A_n will be called adequate if for every sequence $\{s_n\}$, $s_n \in S_n$ such that $\{A_n(s_n)\}$ stays in a precompact subset of V the identities $A_n(s_n) \equiv A_n(t_n)$ imply

$$\lim_n \|P_{s_n,n} - P_{t_n,n}\| = 0.$$

If the sequence $\{A_n\}$ is adequate, one can form experiments $\mathfrak{J}_n = \{F_{v,n}; v \in B_n\}$ with $B_n = A_n(S_n)$ and with $F_{v,n}$ taken equal to one of the $P_{s,n}$ for an index s such that $v = A_n(s)$.

Definition 3. Let the Hilbert space $(V, \|\circ\|)$ be fixed and let $\{B_n\}$ be a sequence of subsets of V. Let $\mathfrak{J}'_n = \{F'_{v,n}; v \in B_n\}$ be two experiments indexed by B_n. The sequences $\{\mathfrak{J}_n\}$, $\{\mathfrak{J}'_n\}$ will be called asymptotically equivalent if for every precompact set $K \subset V$ the distance $\Delta(\mathfrak{J}_{n,K}, \mathfrak{J}'_{n,K})$ between the experiments $\mathfrak{J}_{n,K} = \{F_{v,n}; v \in B_n \cap K\}$ and $\mathfrak{J}'_{n,K} = \{F'_{v,n}; v \in B_n \cap K\}$ tends to zero as $n \to \infty$.

According to these definitions, if A_n is adequate, arbitrary choices of the points s in the sets $A_n^{-1}(v)$ will lead to asymptotically equivalent experiments.

Definition 4. Let $\mathcal{E}_n = \{P_{s,n}; s \in S_n\}$ and let $\{A_n\}$ be an adequate sequence of maps into the fixed Hilbert space $(V, \|\circ\|)$. The sequence $\{\mathcal{E}_n\}$ will be called asymptotically standard Gaussian for the centerings A_n if the corresponding experiments $\mathcal{J}_n = \{F_{v,n}; v \in B_n\}$ are asymptotically equivalent to the standard Gaussian experiment on $(V, \|\circ\|)$.

Definition 5. Let $\{B_n\}$ be a sequence of subsets of $(V, \|\circ\|)$. We shall say that $\{B_n\}$ fills V asymptotically if for every compact $K \subset V$ there are subsets $K_n \subset B_n$ such that the Hausdorff distance between K_n and K tends to zero as $n \to \infty$.

(This definition is related to the concept of tangent cone described in [3]. I have used this concept for some twenty years without giving any references. A closely related concept occurs in the paper [9] of H. Chernoff. I had overlooked the paper and owe the reference to Mr. Moussatat.)

Consider then experiments $\mathcal{E}_n = \{P_{s,n}; s \in S_n\}$ and maps A_n from S_n to the given Hilbert space $(V, \|\circ\|)$. Assume that there is for each n a particular $\theta_n \in S_n$ such that $A_n(\theta_n) = 0$ and define logarithms of likelihood ratios

$$\Lambda_n(t) = \log \frac{d P_{t,n}}{d P_{\theta_n,n}},$$

considered as random variables for the distribution induced by $P_{\theta_n,n}$. Finally, let

$$L_n(t) = \Lambda_n(t) + \tfrac{1}{2}\|A_n(t)\|^2.$$

With these definitions and notations one can state the following.

Theorem 3. Let $\mathcal{E}_n$ and A_n be as described and assume that the images $B_n = A_n(S_n)$ fill V asymptotically. The following statements are equivalent:

(a) <u>The sequence $\{A_n\}$ is adequate and $\{\mathcal{E}_n\}$ is asymptotically standard Gaussian for the centerings</u> A_n.

(b) <u>For all triplets</u> $(s_n, t_n, u_n) \in S_n \times S_n \times S_n$ <u>such that</u> $\{A_n(s_n)\}$, $\{A_n(t_n)\}$ <u>and</u> $\{A_n(u_n)\}$ <u>are precompact sequences of</u> $(V, \|\circ\|)$ <u>the following holds</u>:

   (i) <u>the sequences</u> $\{P_{s_n,n}\}$, $\{P_{t_n,n}\}$ <u>are contiguous</u>,

  (ii) <u>if</u> $A_n(u_n) - [A_n(s_n) + A_n(t_n)] \to 0$ <u>then</u> $L_n(u_n) - [L_n(s_n) + L_n(t_n)]$ <u>tends to zero in probability</u>.

<u>Proof</u>. That (a) ⇒ (b) is easy. To prove that (b) ⇒ (a) one can proceed as follows:

Let u_n and s_n be such that $A_n(u_n) - A_n(s_n) \to 0$. Take t_n equal to the particular element θ_n such that $A_n(\theta_n) = 0$. Then by (b,ii) the difference $L_n(u_n) - L_n(s_n)$ tends to zero in probability for the measures $\{P_{\theta_n,n}\}$. According to the contiguity assumption (b,i), this implies that $\|P_{u_n,n} - P_{s_n,n}\| \to 0$. Thus $\{A_n\}$ is certainly adequate.

Construct experiments $\mathcal{J}_n = \{Q_{v,n}; v \in V\}$ as follows. If $v \in B_n$, let $Q_{v,n}$ be one of the measures $P_{s,n}$, with $A_n(s) = v$. For $v \notin B_n$ take a $w \in B_n$ such that

$$\|w-v\| \leq \frac{1}{n} + \inf\{\|w-u\|; u \in B_n\}$$

and take for $Q_{v,n}$ the measure $Q_{w,n}$ assigned to w. Let M_n be the process defined by the measure $Q_{0,n} = P_{\theta_n,n}$ and the functions

$$M_n(v) = \log \frac{dQ_{v,n}}{dQ_{0,n}} + \tfrac{1}{2}\|v\|^2.$$

Then if the sequences $v_n^{(i)}$, $i=1,2,3$ are precompact and such that $v_n^{(3)} - [v_n^{(1)} + v_n^{(2)}] \to 0$ the difference $M_n[v_n^{(3)}] - [M_n[v_n^{(1)}] + M_n[v_n^{(2)}]]$ will tend to zero in probability. Let $\mathfrak{J} = \{Q_v; v \in V\}$ be a weak cluster point of the sequence of experiments $\{\mathfrak{J}_n\}$. Let Λ be the log-likelihood ratio process taken for Q_0 and let $L(v) = \Lambda(v) + \tfrac{1}{2}\|v\|^2$. Since weak convergence of experiments and the contiguity assumption imply convergence of the finite dimensional distributions of the logarithms of likelihood ratios, one concludes that $L(v+w) = L(v) + L(w)$ almost surely. Thus, by Lemma 3, the cluster point $\mathfrak{J}$ is the standard Gaussian experiment on V. It is uniquely determined. Therefore the $\mathfrak{J}_n$ converge weakly to $\mathfrak{J}$. Finally fix two precompact sequences $\{v_n\}$ and $\{w_n\}$ and let $\mathcal{P}_n$ be the binary experiment $\mathcal{P}_n = (Q_{v_n,n}, Q_{w_n,n})$. Let $\mathcal{P}_n' = (Q_{v_n}, Q_{w_n})$ be taken from the limit $\mathfrak{J}$. The same argument, applied to the linear space spanned by (v_n, w_n) shows that $\Delta(\mathcal{P}_n, \mathcal{P}_n') \to 0$. Thus, according to a theorem of D. Lindae [10], the sequence $\mathfrak{J}_n$ converges to $\mathfrak{J}$ uniformly on the precompact subsets of V. This concludes the proof of the theorem.

The preceding Theorem 3 is presumably applicable to a variety of different situations, but it needs some commentary.

Many authors have considered the case where the experiments $\mathcal{E}_n = \{P_{s,n}; s \in S_n\}$ are direct products of experiments $\mathcal{E}_{n,j} = \{P_{s,j,n}; s \in S_n\}$. Writing $\theta_n = 0$ for simplicity, let us suppose that when $\frac{1}{2}\|P_{s_n,n} - P_{0,n}\|$ stays bounded away from unity, the individual components $\|p_{s,j,n} - p_{0,j,n}\|$ tend to zero uniformly in j. It is then easy to write conditions under which binary experiments $\mathcal{P}_n = (P_{s_n,n}, P_{0,n})$ are asymptotically equivalent to certain binary Gaussian experiments $(G_{s_n,n}, G_{0,n})$. For instance, one may define random variables $Y_{n,j} \geq 1$ so that $(1 + Y_{n,j})^2$ is the density of $p_{s_n,j,n}$ with respect to $p_{0,j,n}$ with the distribution induced by $p_{0,j,n}$. The asymptotically Gaussian behavior of the pairs is then implied by the requirements that (i) the $Y_{n,j}$ satisfy Lindberg's condition and (ii) if $\beta_{n,j}$ is the mass of $p_{s_n,j,n}$ which is $p_{0,j,n}$ singular, then $\Sigma_j \beta_{n,j} \to 0$. In fact, these two conditions are equivalent to the contiguity and asymptotic normality of the sequences $\{(P_{s,n}, P_{0,n})\}$. The linearity property involved in Theorem 3 can then be checked by taking limits of covariances. (See [11].)

The classical studies often associated with the theory of maximum likelihood estimates revolves around structures where there is a set Θ, open subset of a Euclidean space R^k. The Hilbert space V is also a Euclidean space R^k. The sets S_n are neighborhoods of a given point $\theta_0 \equiv \theta_n$ and the maps A_n are linear. However, even if one limits oneself to independent identically distributed observations, there is no lack of examples where the Hilbert space

V is infinite dimensional. One case, described by Prakasa Rao [12], is one in which Θ is the line. The observations are drawn from a shift family $f(x - \theta)$, $\theta \in \Theta$. The S_n are neighborhoods of a given point $\theta_0 = 0$ and $P_{s,n}$ is the distribution of n independent observations from a density $f(x - \delta_n s)$, for a suitable constant δ_n.

If f has a cusp of the type $\exp\{-|x|^\alpha\}$ with $0 < \alpha < \frac{1}{2}$, the corresponding Hilbert space V is infinite dimensional, even though the families $\{P_{s,n}\}$ have a bounded dimension in the covering sense of Sections 2,5 and 6.

Another example is the one treated by D. M. Chibisov [13]. There V is the space of functions s defined on [0,1] and such that $\int s d\lambda = 0$ and $\int s^2 d\lambda < \infty$ for the Lebesgue measure λ. The measure $P_{s,n}$ is the joint distribution of n independent observations from the measure whose density with respect to λ is $f_{s,n}(x) = 1 + 1/\sqrt{n}\ s(x)$.

Of course, Theorem 3 can in principle be applied to cases where the observations are not independent, but the specific results presently available are far from satisfactory.

To give a possible example, consider a situation where $\mathcal{E}_n$ consists in observing a sequence $\{X_{j,n};\ j=1,2,\ldots\}$ such that the conditional distributions $\mathcal{L}[X_{j+1,n}|X_{n,i},\ i=1,2,\ldots,j]$ are given by Markov kernels $F_{s,n}[x_1, x_2, \ldots, x_j; B]$. For fixed $x_1, x_2, \ldots, x_j$ let $1 + Z_{n,j+1}(s; x_1, \ldots, x_j)$ be the density of the measure $F_{s,n}$ with respect to $F_{0,n}$. This density contains a free variable x_{j+1} which we shall take distributed according to

$F_{0,n}(x_1, x_2, \ldots, x_j; dx_{j+1})$. With these notations one obtains conditional log-likelihoods $\Lambda_{n,j+1}[s; x_1, \ldots, x_j]$. The combined log-likelihood process is their sum $\Lambda_n(s) = \Sigma_j \Lambda_{n,j}(s)$.

One can then obtain asymptotic normality results by direct and brutal application of the theorems of A. Dvoretzky [5].

Another possiblity is suggested by a recent theorem of D. L. McLeish [6].

Consider not the logarithms $\Lambda_{n,j}$ but the processes $Z_{n,j} = \exp\{\Lambda_{n,j}\} - 1$ defined above. Fix a sequence $\{s_n\}$ and write $Z_{n,j} = Z_{n,j}(s_n; x_1, \ldots, x_{j-1})$ for short.

In his theorem 2.3 McLeish uses conditions (2.3) (b) and (2.3) (c) as follows:

(b) $\max_j |Z_{n,j}| \to 0$ in probability.

(c) $\Sigma_j Z_{n,j}^2 \to \sigma^2$ in probability.

The usual Taylor expansion argument shows that, writing $\Lambda_n = \Sigma_j \Lambda_{n,j}$ the difference $\Lambda_n - \Sigma_j Z_{n,j}$ tends to $-\tfrac{1}{2}\sigma^2$ in probability.

One should note that in the independent case and with individually negligible contributions, both of McLeish conditions (b) and (c) are necessary. Either one of them is also sufficient to imply that the distributions of logarithms of likelihood ratio are approximately Gaussian. They do not imply contiguity.

If the conditional distributions $F_{s_n,n}(x_1, \ldots, x_j; B)$ are absolutely continuous with respect to $F_{0,n}(x_1, \ldots, x_j; B)$ the $Z_{n,j}$ are martingale differences. It follows then from McLeish's theorem that $\mathcal{L}(\Lambda_n|P_{0,n}) \to n(-\frac{1}{2}\sigma^2, \sigma^2)$ whenever $\max_j |Z_{n,j}|$ stays uniformly bounded in L_2-norm. This will imply the contiguity conditions of Theorem 3.

Another possibility (under McLeish conditions (2.3)(b)(c)) is to check that for triplets (s_n, t_n, u_n) satisfying the requirements $A_n(u_n) - [A_n(s_n) + A_n(t_n)] \to 0$ of part (b,ii) of Theorem 3 the difference

$$\sum_j \{Z_{n,j}(u_n) - [Z_{n,j}(s_n) + Z_{n,j}(t_n)]\}$$

tends to zero in probability. To conclude it is then sufficient to prove the appropriate contiguity statements.

To terminate, we wish to point out that even if one can obtain asymptotic normality statements in this manner, they will only be local statements valid in small shrinking neighborhoods of particular values of the parameter θ.

To obtain a theorem similar to the combination of Theorems 4 and 5 of [3], Section 12, it remains necessary to construct estimates having properties analogous to those used in Sections 3 and 4. We have been unable to extend the construction of Section 4 to the dependent case, but recent results of I. Zhurbenko [14] suggest that a construction will be possible in cases where the dependence is not too far from a Markov dependence.

REFERENCES

[1] A. Wald. "Tests of statistical hypotheses concerning several parameters when the number of observations is large." <u>Trans. Amer. Soc.</u>, Vol. 54 (1943), pp. 426-482.

[2] L. Le Cam. "On the asymptotic theory of estimation and testing hypotheses." <u>Proc. Third Berkeley Symp. on Math. Stat. and Prob.</u> Vol. 1 (1956), pp. 129-156.

[3] L. Le Cam. <u>Notes on asymptotic methods in decision theory.</u> Chapter I. Publications du Centre de Recherches Mathématiques, Université de Montréal (1973) xvi + 270 pages.

[4] L. Le Cam. "On the information contained in additional observations." <u>Ann. Statist.</u> Vol. 2 (1974), pp. 630-649.

[5] A. Dvoretzky. "Asymptotic normality of sums of dependent random variables." <u>Proc. Sixth Berkeley Symp. on Math. Stat. and Prob.</u> Vol. II (1972), pp. 513-535.

[6] D. M. McLeish. "Dependent central limit theorems and invariance principles." <u>Ann. Prob.</u> Vol. 2 (1974), pp. 620-628.

[7] A. N. Kolmogorov. "On some asymptotic characteristics of totally bounded metric spaces." <u>Doklady Akad Nauk SSSR.</u> Vol. 108 (3) (1956).

[8] I. A. Ibragimov and R. Z. Has'minskii. "Asymptotic behavior of statistical estimates for samples with a discontinuous density." <u>Math. USSR Sbornik.</u> Vol. 16 (1972), pp. 573-606.

[9] H. Chernoff. "On the distribution of the likelihood ratio." <u>Ann. Math. Statist.</u> Vol. 25 (1954), pp. 573-578.

[10] D. Lindae. "Distributions of likelihood ratios and convergence of experiments." Ph.D. Thesis, University of California, Berkeley (1972) (unpublished).

[11] L. Le Cam. <u>Théorie asymptotique de la décision statistique.</u> Les Presses de l'Université de Montréal (1969), 140 pages.

[12] B. L. S. Prakasa Rao. "Estimation of the location of the cusp of a continuous density." <u>Ann. Math. Statist.</u> Vol. 39 (1968), pp. 76-87.

[13] D. M. Chibisov. "Transition to the limiting process for deriving asymptotically optimal test." <u>Sankhyā, Series A.</u> Vol. 31 (1969), pp. 241-258.

[14] I. Zhurbenko. "Mixing conditions and asymptotic theory of stationary time series." (To be published.)

Statistical Inference for
Stationary Point Processes[1]

by David R. Brillinger

The University of California, Berkeley

Introduction

This work is divided into three principal sections which also correspond to the three lectures given at Bloomington. The topics cover, some useful point process parameters and their properties, estimation of time domain parameters and the estimation of frequence domain parameters. The work may be viewed as an extension of some of the results in Cox and Lewis (1966, 1972) to apply to vector-valued processes and to higher order parameters. It will proceed at a heuristic level rather than formal. A formal approach may be found in Daley and Vere-Jones (1972) for example. The notation $\int f$ will be used for $\int f(x)\, d\mu(x)$, μ being Lebesgue measure. A general lemma concerning the existence of consistent estimates is given in Section IV.

I. Point Process Parameters

Consider isolated points of r different types randomly distributed along the real line, R.

---

[1] Prepared while the author was a Miller Research Professor and with the support of N.S.F. Grant GP-31411.

Examples that we have in mind include, the times of heart beats or earthquakes in the case r = 1, the times of nerve pulses released by a network of r nerve cells in the case of general r. Let $N_a(A)$ denote the number of points of type a falling in the interval A ⊂ R and let $N_a(t) = N_a(0,t]$ for a = 1,...,r.

1. Suppose

Prob {point of type a in (t, t+h]} ~ $p_a(t)h$
as h ↓ 0 . $p_a(t)$ provides a measure of the <u>intensity</u> with which points of type a occur near t. We can often conclude that

$$E\ N_a(t) = \int_0^t p_a(t)\ dt$$

2. Suppose, for $t_1 \neq t_2$

Prob{point of type a in $(t_1, t_1+h_1]$ and point of type b in $(t_2, t_2+h_2]$}

$$\sim p_{ab}(t_1, t_2) h_1 h_2$$

as h_1, h_2 ↓ 0. $p_{ab}(t_1, t_2)$ provides a measure of the intensity with which points of type a occur near t_1 and simultaneously points of type b occur near t_2.

A related useful measure is provided by

Prob{point of type a in $(t_1, t_1+h]$ | point of type b at t_2}

$$\sim p_{ab}(t_1, t_2) h / p_b(t_2)$$

as $h \downarrow 0$. The ratio $p_{ab}(t_1,t_2)/p_b(t_2)$ is seen to provide a measure of the intensity with which type a points occur near t_1, given that there is a type b point at t_2. In the case that type a points are distributed independently of type b points, $p_{ab}(t_1,t_2) = p_a(t_1)p_b(t_2)$, and the ratio becomes $p_a(t_1)$, the first order intensity. The function $p_{ab}(t_1,t_2)$ is like the second order moment function of ordinary time series; however in practise it seems to be much more useful as it has a further interpretation as a probability. Often it is true that

$$E\, N_a(t)\, N_b(t) = \int_0^t \int_0^t p_{ab}(t_1,t_2) dt_1 dt_2 \quad \text{for } a \neq b$$

$$= \int_0^t \int_0^t p_{ab}(t_1,t_2) dt_1 dt_2 + \int_0^t p_a(t)\, dt \quad \text{for } a = b$$

3. Suppose next that, for $t_1,\ldots,t_k$ distinct and $\nu_1,\ldots,\nu_r$ non-negative integers with sum k

Prob{type a point in **each** of $(t_j, t_j+h_j]$,

$j = \sum_{b<a} \nu_b + 1, \ldots, \sum_{b \leq a} \nu_b$ and $a = 1,\ldots,r\}$

$\sim p^{(\nu_1)\cdots(\nu_r)}(t_1,\ldots,t_k)h_1\cdots h_k$ \hfill (1)

as $h_1,\ldots,h_k \downarrow 0$; $k = 1,2,\ldots$. (An alternate notation, consistent with the cases k = 1,2 above is

$$\text{Prob}\{\text{type } a_j \text{ point in } (t_j, t_j+h_j], \ j = 1,\ldots,k\}$$

$$\sim p_{a_1\cdots a_k}(t_1,\ldots,t_k) \ h_1\cdots h_k$$

as $h_1,\ldots,h_k \downarrow 0$; $k = 1,2,\ldots$.) The function $p^{(\nu_1)\cdots(\nu_r)}$ is called a <u>product density of order k</u>. Such a function was introduced by S. O. Rice in a particular situation and by A. Ramakrishnan in a general situation, see Srinivasan (1974). No claim is made that the probability in (1) always depends on $h_1,\ldots,h_k$ in such a direct manner. Rather it is the claim that this happens for an interesting class of examples. Brillinger (1972) gives an expression for

$$E \ N_{a_1}(t_1) \cdots N_{a_k}(t_k)$$

4. The <u>probability generating functional</u> of the process $\underset{\sim}{N}(t) = \{N_1(t),\ldots,N_r(t)\}$ is defined by

$$G[\xi_1,\ldots,\xi_r] = E[\exp\{\int \log \xi_1(t) \ dN_1(t) + \cdots + \int \log \xi_r(t) \ dN_r(t)\}]$$

for suitable functions $\xi_1\cdots\xi_r$. Writing it as

$$E[\prod_{a=1}^{r} \prod_{\tau \text{ type } a \text{ point}} \{1 + (\xi_a(\tau)-1)\}]$$

and expanding, we can see that it is given by

$$\sum_{\nu_1,\ldots,\nu_r} \frac{1}{\nu_1!\cdots\nu_r!} \int (\xi_1-1)^{(\nu_1)} \cdots (\xi_r-1)^{(\nu_r)} \ p^{(\nu_1)\cdots(\nu_r)}.$$

where we define

$$(\xi(t_1,\ldots,t_\nu) - 1)^{(\nu)} = (\xi(t_1) - 1)\ldots(\xi(t_\nu)-1)$$

This functional is of use in computing probabilities of interest for the process. For example setting

$$\xi_a(t) = z_a \quad \text{for } t \in A$$
$$= 1 \quad \text{for } t \notin A$$

and determining the coefficient of $z_1^{j_1}\ldots z_r^{j_r}$ we see that

$$\text{Prob}\{N_1(A) = j_1,\ldots,N_r(A) = j_r\}$$

$$= \frac{1}{j_1!\ldots j_r!} \sum_{\nu_1 \geq j_1} \ldots \sum_{\nu_r \geq j_r} \cdot$$

$$\frac{(-1)^{\nu_1-j_1 + \ldots + \nu_r-j_r}}{(\nu_1-j_1)!\ldots(\nu_r-j_r)!} \cdot$$

$$\int_A p^{(\nu_1)\ldots(\nu_r)}_{\nu_1+\ldots+\nu_r} \qquad (2)$$

We may likewise determine <u>conditional product densities</u> such as

$$p^{(\nu_1)\ldots(\nu_r)}(t_1,\ldots,t_k \mid N_1(A) = j_1,\ldots,N_r(A) = j_r)$$

$$= \frac{1}{(j_1-\nu_1)!\ldots(j_r-\nu_r)!} \sum_{\nu_1 \geq j_1} \ldots \sum_{\nu_r \geq j_r}$$

$$\frac{(-1)^{\nu_1-j_1+\ldots+\nu_r-j_r}}{(\nu_1-j_1)!\ldots(\nu_r-j_r)!} \int_A \nu_1-\gamma_1+\ldots+\nu_r-\gamma_r$$

$$p^{(\nu_1)\ldots(\nu_r)}(t_1,\ldots,t_{\gamma_1};\ldots;t_{\gamma_1+1},\ldots,t_{\gamma_1+\gamma_2};\ldots;$$

$$\ldots)/(2)$$

These conditional product densities are useful in statistical inference. They provide likelihood functions and also allow the investigation of the distribution of statistics conditionally on the observed number of points. (Were $N(A) = 0$, one wouldn't want to claim much.)

The integrated product densities give the factorial moments of the process. For example, if $N_{(\nu)} = N(N-1)\ldots(N-\nu+1)$, then

$$E\, N_1(A)_{(\nu_1)}\ldots N_r(A)_{(\nu_r)} = \int_{A^k} p^{(\nu_1)\ldots(\nu_r)}$$

Also of use are the <u>cumulant densities</u>, $q^{(\nu_1)\ldots(\nu_r)}(t_1,\ldots,t_k)$ given by

$$\log G[\xi_1,\ldots,\xi_r] = \sum_{\nu_1,\ldots,\nu_r} \frac{1}{\nu_1!\ldots\nu_r!}$$

$$\int (\xi_1-1)^{(\nu_1)}\ldots(\xi_r-1)^{(\nu_r)} q^{(\nu_1)\ldots(\nu_r)} \quad (3)$$

They measure the degree of dependence of increments of the process at different t_j.

Certain other conditional product densities are of use. We mention

Prob{type a point in each of $(t_j, t_j+h_j]$, j = $\sum_{b<a} \nu_b + 1, \ldots, \sum_{b \leq a} \nu_b$ and a = $1, \ldots, r | N_1\{0\} = 1\}/ (h_1 \ldots h_k)$

$$\sim p^{(\nu_1) \ldots (\nu_r)}(t_1, \ldots, t_k \mid N_1\{0\} = 1)$$

$$= p^{(\nu_1+1)(\nu_2) \ldots (\nu_r)}(0, t_1, \ldots, t_k)/p_1(0)$$

and for $\tau_1, \ldots, \tau_k \leq t$

Prob{type 1 point in $(t, t+h]$ | ν_1 points of type 1, ν_2 points of type 2, ... at $\tau_1, \tau_2, \ldots, \tau_k$ respectively}/n

$$\sim p^{(\nu_1+1)(\nu_2) \ldots (\nu_r)}(t, \tau_1, \tau_2, \ldots, \tau_k)/$$

$$p^{(\nu_1) \ldots (\nu_r)}(\tau_1, \ldots, \tau_k)$$

If all points up to t are included, this becomes the <u>complete intensity</u>

$$\lim_{h \downarrow 0} \text{Prob}\{\text{type 1 point in } (t, t+h] \mid \underset{\sim}{N}(u), u \leq t\}$$

5. Certain probabilities and moments are of special interest. We list some of these.

(i) the **renewal functions**

$$U_{ab}(t) = E\{N_a(t) \mid N_b\{0\} = 1\} \quad \text{for } t > 0$$

$$= \int_0^t p_{ab}(u,0) \, du \, / \, p_b(0) \quad a,b=1,\ldots,r.$$

The renewal density is $p_{ab}(t,0)/p_b(0)$.

(ii) the **forward recurrence time distribution** is given by

Prob{event before or at t}
  = Prob{time of next event from 0 is ≤ t}
  = 1 - Prob{N(t) = 0}

$$= 1 - \sum_{\nu \geq 0} \frac{(-1)^\nu}{\nu!} \int_{(0,t]^\nu} p^{(\nu)}$$

(iii) the **survivor function** (or distribution of lifetime)

Prob{time of next event from 0 is > t | N{0} = 1}
  = Prob{N(t) = 0 | N{0} = 1}

$$= p(0)^{-1} \sum_{\nu \geq 0} \frac{(-1)^\nu}{\nu!} \int_{(0,t]^\nu} p^{(\nu+1)}(0,\ldots)$$

  = 1 - F(t) say.

(iv) the **hazard function** or **force of mortality**

$$\mu(t) = f(t)/(1 - F(t))$$

$$\sim \text{Prob}\{\text{point in } (t,t+h] \mid N\{0\} = 1, N(t) = 0\}/h$$

where F(t) is given in (iii) and f(t) is its derivative.

STATISTICAL INTERFERENCE

(v) the <u>variance time curve</u>

$$\text{var } N(t) = E\, N(t)(N(t) - 1) + E\, N(t) - (E\, N(t))^2$$
$$= \int_0^t \int_0^t p^{(2)}(t_1,t_2)dt_1\, dt_2 + \int_0^t p(t)\, dt - (\int_0^t p(t)dt)^2$$

(vi) the <u>Palm functions</u>

$$q_1(j_1,\ldots,j_r\,;\,t) = \text{Prob}\{N_1(t) = j_1,\ldots,N_r(t) = j_r \mid N_1\{0\} = 1\}$$

$$= \frac{1}{j_1!\,\ldots\,j_r!\,p_1(0)} \sum_{\nu_1 \geq j_1} \cdots \sum_{\nu_r \geq j_r} \frac{(-1)^{\nu_1-j_1+\ldots+\nu_r-j_r}}{(\nu_1-j_1)!\,\ldots\,(\nu_r-j_r)!}$$

$$\int_{(0,t]^{\nu_1+\ldots+\nu_r}} p^{(\nu_1+1)(\nu_2)\ldots(\nu_r)}(0,\ldots$$

6. We next indicate the values of a few of these parameters for some examples of interest.

<u>Example 1.</u> The Poisson process with mean intensity $p(t)$. The numbers of points in disjoint intervals $I_1,\ldots,I_k$ are independent Poisson variates with means $P(I_1),\ldots,P(I_k)$ respectively where $P(I) = \int_I p(t)\, dt$. Here

$$p^{(k)}(t_1,\ldots,t_k) = p(t_1)\ldots p(t_k)$$

and so

$$G[\xi] = \exp\{\int (\xi(t) - 1)\, p(t)\, dt\}$$
$$\text{Prob}\{N(A) = j\} = \frac{1}{j!}\, P(A)^j \exp\{-P(A)\}$$
$$p^{(k)}(t_1,\ldots,t_k \mid N(A) = j) = \frac{j!}{(j-k)!\,P(A)}\cdot\frac{p(t_1)}{P(A)}\ldots\frac{p(t_k)}{P(A)}$$

63

If $P(t) = \int_0^t p(t)\, dt$ and $N'(s)$, $s \in R_+$ is a Poisson process with mean intensity 1, then the general process may be represented as

$$N(t) = N'(P(t))$$

Example 2. The doubly stochastic Poisson process. Suppose $\{x_1(t),\ldots,x_r(t)\}$, $t \in R_+$, is a process with non-negative sample paths, moments

$$m^{(\nu_1)\cdots(\nu_r)}(t_1,\ldots,t_k) = E\{x_1(t_1)\cdots x_1(t_{\nu_1}) x_2(t_{\nu_1+1})\cdots x_r(t_k)\}$$

and moment generating functional

$$M[\theta_1,\ldots,\theta_r] = E[\exp\{\int \theta_1(t)x_1(t)\,dt + \cdots + \int \theta_r(t)x_r(t)\,dt\}]$$

Suppose after a realization of this process is obtained, independent Poissons with mean intensities $x_1(t),\ldots,x_r(t)$ are generated. Then

$$p^{(\nu_1)\cdots(\nu_r)}(t_1,\ldots,t_k) = m^{(\nu_1)\cdots(\nu_r)}(t_1,\ldots,t_k)$$
$$G[\xi_1,\ldots,\xi_r] = M[\xi_1-1,\ldots,\xi_r-1]$$
$$= E[\exp\{\int (\xi_1(t)-1)x_1(t)\,dt + \cdots\}]$$

If $X_a(t) = \int_0^t x_a(t)\,dt$, and $N_1'(s),\ldots,N_r'(s)$ are independent Poissons with mean intensities 1, then this process may be represented as

STATISTICAL INTERFERENCE

$$N_1'(X_1(t)),\ldots,N_r'(X_r(t)) \;.$$

This process seems to be useful for checking out general formulas that have been developed, such as (2) and (3), among other things.

Example 3. The cluster process. Suppose $N_1'(t),\ldots,$ $N_r'(t)$ is a primary process of cluster centers with probability generating functional $G[\xi_1,\ldots,\xi_r]$. Suppose that secondary points are generated in independent clusters centered at the points of $\underset{\sim}{N}'$. Suppose that the p.g.f. for cluster points of type a centered at t is $G_a[\xi|t]$. Then the p.g.f. of the overall process is

$$G[\xi_1,\ldots,\xi_r] = E\{\prod_{j,k}\xi_1[\sigma_j^1+\tau_{jk}^1]\ldots\prod_{j,k}\xi_r[\sigma_j^r+\tau_{jk}^r]\}$$
$$= E\{\prod_j G_1[\xi_1|\sigma_j^1]\ldots\prod_j G_r[\xi_r|\sigma_j^r]\}$$
$$= G[G_1[\xi_1|\cdot],\ldots,G_r[\xi_r|\cdot]]$$

If $r = 2$, and the first component is the primary process and the second component corresponds to clusters of one member, then we have a process of the character of the G/G/∞ queue.

Example 4. The renewal process. Here the points correspond to the partial sums of a random walk with positive steps. Suppose $r = 1$, $t_1 < t_2 < \ldots < t_k$, then

$$p^{(k)}(t_1,\ldots,t_k) = p^{(1)}(t_1)\frac{p^{(2)}(t_2,t_1)}{p^{(1)}(t_1)}\frac{p^{(2)}(t_3,t_2)}{p^{(1)}(t_2)}$$

$$\cdots \frac{p^{(2)}(t_k,t_{k-1})}{p^{(1)}(t_{k-1})}$$

where $p^{(1)}$ and $p^{(2)}$ satisfy renewal equations, see p. 35 in Srinivasan (1974).

Example 5. Zero crossing processes. Expressions may be set down for the product densities of point processes corresponding to the zeros of random functions, see Leadbetter (1972).

7. We now turn to a consideration of the case in which the process is stationary, that is, probability distributions are invariant under translations of t. This means for example,

$$p_a(t) = p_a$$
$$p_{ab}(t_1,t_2) = p_{ab}(t_1-t_2)$$
$$p^{(\nu_1)\ldots(\nu_r)}(t_1,\ldots,t_k) = p^{(\nu_1)\ldots(\nu_r)}(t_1-t_k,\ldots,t_{k-1}-t_k)$$

and if S^u denotes the shift transformation, $S^u \xi(t) = \xi(t+u)$, then

$$G[S^u \xi_1,\ldots,S^u \xi_r] = G[\xi_1,\ldots,\xi_r].$$

As the process has stationary increments, it has a spectral representation

$$N_a(t) = \int_{-\infty}^{\infty} [(\exp\{it\lambda\}-1)/(i\lambda)] \, dZ_a(\lambda)$$

for $a = 1,\ldots,r$. We may define <u>cumulant spectra</u> of order k by

$$\delta(\lambda_1+\ldots+\lambda_k) \, f^{(\nu_1)\ldots(\nu_r)}(\lambda_1,\ldots,\lambda_{k-1}) d\lambda_1 \ldots d\lambda_k$$
$$= \text{cum}\{dZ_1(\lambda_1),\ldots,dZ_1(\lambda_{\nu_1}),\ldots,dZ_r(\lambda_k)\}$$

with $\delta(\bullet)$ the Dirac delta function. Alternately, making use of product densities, we might define the <u>power spectra</u> by

$$f_{aa}(\lambda) = (2\pi)^{-1}[p_a + \int_{-\infty}^{\infty} \{p_{aa}(u) - p_a^2\}\exp\{-i\lambda u\}d\lambda]$$

$-\infty < \lambda < \infty$, $a = 1,\ldots,r$, since

$$\text{cov}\{dN_a(t+u), dN_a(t)\} = [p_a\delta(u) + \{p_{aa}(u)-p_a^2\}]dt \, du$$

and <u>cross-spectra</u> by

$$f_{ab}(\lambda) = (2\pi)^{-1}\int_{-\infty}^{\infty} \{p_{ab}(u)-p_a p_b\} \exp\{-i\lambda u\}du$$

$-\infty < \lambda < \infty$, $1 \le a \ne b \le r$, since

$$\text{cov}\{dN_a(t+u), dN_b(t)\} = \{p_{ab}(u)-p_a p_b\}dt \, du$$

The functions $q_{ab}(u) = p_{ab}(u) - p_a p_b$ tend to 0 as $|u| \to \infty$ and are integrable for many processes (processes whose distant increments are only weakly dependent.) In this connection we set down the mixing condition,

Assumption I. $\underline{N}(t)$, $t \in R$, is an r vector-valued stationary point process satisfying (1), whose cumulant densities of (3) satisfy

$$\int \cdots \int_{R^{k-1}} |q^{(\nu_1)\cdots(\nu_r)}(u_1,\ldots,u_{k-1})| du_1 \cdots du_{k-1} < \infty$$

for $\nu_1,\ldots,\nu_r = 0,1,2,\ldots,\nu_1 + \cdots + \nu_r \geq 2$.

The second-order spectra of the process, $f_{ab}(\lambda)$, possess many of the same properties as the spectra of ordinary time series. There are however some differences, we mention that

$$\lim_{|\lambda| \to \infty} f_{aa}(\lambda) = (2\pi)^{-1} p_a$$

for mixing point processes instead of the 0 limit for mixing ordinary time series.

The spectral representation may be used to relate the point process to the associated ordinary time series

$$\underline{X}(t) = h^{-1}\underline{N}(t-\tfrac{h}{2}, t+\tfrac{h}{2}) = \int \exp\{i\lambda t\}[(\sin h\lambda/2)/(h\lambda/2)] d\underline{Z}(\lambda)$$

$t \in R$. This shows, for example, that the cross-spectrum of the a-th and b-th components of $\underline{X}(t)$ is

$$[(\sin h\lambda/2)/(h\lambda/2)]^2 f_{ab}(\lambda)$$

8. A key indicator of the appearance of the process of points of type 1, say, is provided by

$$h^{-1} N_1(t,t+h] \qquad h \text{ small}$$

the empirical intensity with which points of type 1 are seen to occur near t. Models for the process may usefully involve models for this variate. A simple statement says

$$\text{Prob}\{\text{point of type 1 in } (t,t+h]\} \sim p_1 h$$

for h small. A more complicated statement is

$$\text{Prob}\{\text{point of type 1 in } (t,t+h] \mid \text{point of type } a \text{ at } \tau\}$$

$$\sim p_{1a}(t-\tau)h/p_a$$

In the case that the process 1, near t, is independent of the process a, near τ, this last is $\sim p_1 h$, the marginal intensity. This happens often as $|t-\tau| \to \infty$. An even more complicated statement involves

Prob{point of type 1 in $(t,t+h]$ | ν_1 points of type 1, ν_2 points of type 2, ... at $\tau_1, \tau_2, \ldots, \tau_k$ respectively}

$$\sim p^{(\nu_1+1)(\nu_2)\ldots(\nu_r)}(t-\tau_k, \tau_1-\tau_k, \ldots, \tau_{k-1}-\tau_k)h/$$
$$p^{(\nu_1)\ldots(\nu_r)}(\tau_1-\tau_k, \ldots, \tau_{k-1}-\tau_k)$$

Suppose $r = 2$. A useful simple model here is;

Prob{point of type 1 in $(t,t+h]$ | $N_2(u)$, $-\infty < u < \infty$}

$$\sim \{\mu + \int a(t-u)\, dN_2(u)\}h \qquad (4)$$

$$\sim \{\mu + \sum_j a(t-\tau_j)\}h$$

where the τ_j are the times of the events of the second process. This model allows the intensity, near t, of points of type 1 to be affected in a direct manner by points of type 2. If the system is causal, then $a(u) = 0$, $u < 0$. The second process may excite or inhibit the first process depending on the sign of $a(u)$.

The model implies, for example,

$$p_1 = \mu + p_2 \int a(u)\, du \qquad (5)$$

showing that μ may be interpreted as the intensity with which type 1 points would occur where $p_2 = 0$. Also

$$p_{12}(t) = \mu p_2 + p_2 a(t) + \int a(t-u)\, p_{22}(u)\, du \qquad (6)$$

If

$$A(\lambda) = \int a(u)\, \exp\{-i\lambda u\}\, du$$

then (5) and (6) lead to

$$p_1 = \mu + p_2\, A(0)$$

$$f_{12}(\lambda) = A(\lambda)\, f_{22}(\lambda)$$

suggesting how the parameters μ, $A(\lambda)$ might be identified. If $p_{22}(u)$ is constant, as in the Poisson case, then (6) leads to

$$p_{12}(t)/p_2 = c + a(t)$$

and $a(t)$ may be measured directly.

As an example of the model (4) we mention the $G/G/\infty$ queue with N_1 referring to the process of exit times, N_2 to the process of entry times, $a(-u)$ referring to the density of service times and $\mu = 0$. Clearly, here

Prob{customer leaves in the interval $(t,t+h]$ | $N_2(u)$, $-\infty < u < \infty$}

$$\sim \{\sum_j a(t-\tau_j)\}h$$

An interesting problem is that of measuring the degree of association of two point processes. A measure suggested by the preceding model is the <u>coherence</u>

$$|R_{12}(\lambda)|^2 = |f_{12}(\lambda)|^2/(f_{11}(\lambda)f_{22}(\lambda))$$

see Brillinger (1974a). This parameter also appears as a measure of the degree of linear predictability of the process N_1 by the process N_2. It satisfies $0 \leq |R_{12}(\lambda)|^2 \leq 1$. Other measures of association could be based on the nearness of the function $p_{12}(u) - p_1 p_2$ to 0.

We mention next the <u>self-exciting processes</u> introduced by Hawkes, see Hawkes (1972). For

$r = 1$, these satisfy

$$\text{Prob}\{\text{point in } (t, t+h] \mid N(u), u \leq t\}$$
$$\sim \{\mu + \int_{-\infty}^{t} a(t-u) \, dN(u)\}h$$
$$\sim \{\mu + \sum_{\tau_j \leq t} a(t-\tau_j)\}h$$

If we have more than one process, then we could also set up multivariate linear models and define partial parameters. As another extension, we could consider non-linear models such as

$$\text{Prob}\{\text{point of type 1 in } (t, t+h] \mid N_2(u), -\infty < u < \infty\}$$
$$\sim \{a_0 + \int a_1(t-u) \, dN_2(u) + \iint_{u \neq v} a_2(t-u, t-v) \, dN_2(u)$$
$$dN_2(v)\}h \quad .$$

More details concerning such extensions may be found in Brillinger (1974b).

9. We end by mentioning that some, possibly unexpected, relationships exist between certain of the parameters that have been defined. These are the <u>Palm-Khinchin relations</u>,

$$\text{Prob}\{N(t) \leq j\} = p \int_{t}^{\infty} \text{Prob}\{N(u) = j \mid N\{0\} = 1\} du$$
$$= 1 - p \int_{0}^{t} \text{Prob}\{N(u) = j \mid N\{0\} = 1\} du$$
$$\text{Prob}\{N(t) > j \mid N\{0\} = 1\} = 1 + D^{+}\{p^{-1} \sum_{j=0}^{j} (j+1-k) \cdot$$
$$\text{Prob}\{N(t) = k\}\}$$
$$E\{N(t)(N(t)-1) \ldots (N(t) - k)\} =$$

$$= (k+1) \, p \int_0^t E\{N(u) \, (N(u) - 1) \, \ldots \, (N(u) - k + 1) \mid N\{0\} = 1\} \, du$$

Such relationships are discussed in Cramér, Leadbetter and Serfling (1971).

In this first section of the paper we have sought to provide a framework within which stationary point processes may be handled when the only element of statistical independence is asymptotic.

II. Estimation of Time Domain Parameters for Stationary Processes

We consider the estimation of certain time domain parameters given a realization of a process $\underline{N}(t)$ over the interval $(0,T]$, i.e. given the observed times of events in $(0,T]$. We begin with the first order mean intensities p_a, $a = 1,\ldots,r$.

1. Obvious estimates of the p_a, $a = 1,\ldots,r$, are the

$$\hat{p}_a = N_a(T)/T$$

$a = 1,\ldots,r$. In connection with these we have,

Theorem 1. Suppose the process satisfies Assumption I. Then $[\hat{p}_1,\ldots,\hat{p}_r]$ is asymptotically

$$N_r([p_1,\ldots,p_r] \, ; \, 2\pi \, T^{-1}[f_{ab}(0)])$$

as $T \to \infty$.

This theorem, as are those given later, is proved in the final section of the paper. The estimates are asymptotically normal. The asymptotic variance of $\hat{p}_a$ is $2\pi\, T^{-1} f_{aa}(0)$. Were increments of the process uncorrelated, this would be $T^{-1} p_a$. We will see how to estimate $f_{aa}(\lambda)$ next section. Were T large, we might set $T = JU$ and take

$$(TU(J-1))^{-1} \sum_{j=0}^{J-1} (N_a(jU,(j+1)U] - N(0,T]/J)^2 \sim$$
$$2\pi\, T^{-1} f_{aa}(0) \chi^2_{J-1}/(J-1) .$$

The ratio $2\pi f_{aa}(0)/p_a$ is useful in describing certain aspects of the process N_a. If it is greater than 1, the process is said to be <u>clustered</u> or <u>underdispersed</u>. If it is less than 1, the process is called <u>overdispersed</u>.

2. In the second order case we are interested in estimating

$p_{ab}(u) \sim \text{Prob}\{\text{type a in } (t+u,t+u+h_1] \text{ and type b in } (t,t+h_2]\}/(h_1 h_2)$ for $u \neq 0$ and

$p_{ab}(u)/p_b \sim \text{Prob}\{\text{type a in } (t+u,t+u+h] \mid \text{type b at } t\}/h$ for $u \neq 0$.

It seems natural to base estimates of these on

$$J_{ab}^T(u) = \#\{(j,k) \text{ such that } u - \beta < \tau_j^a - \tau_k^b < u + \beta$$
$$\text{and } \tau_j^a \neq \tau_k^b\} \qquad (7)$$

for some small bin width $2\beta > 0$. On the CDC 6400, this statistic can be computed about twice as fast as a direct convolution based on $N(T)$ values.

In connection with this variate we have,

Theorem 2. Suppose the process $\underline{N}$ satisfies Assumption I and that $p_{ab}(\cdot)$ is a continuous function for $a,b = 1,\ldots,r$. Suppose $J_{ab}^T(u)$ is given by (7) with $\beta = \beta_T$ depending on T. Suppose $u_k^T \to u_k$ with $|u_k^T - u_{k'}^T|^2 \geq 2\beta_T$ for $1 \leq k < k' \leq K$. Then as $T \to \infty$, (i) if $\beta_T = L/T$, L fixed, the variates $J_{a_1 b_1}^T(u_1^T),\ldots,J_{a_K b_K}^T(u_K^T)$ are asymptotically independent Poissons with means $2\beta_T T\, p_{a_k b_k}(u_k)$, $k = 1,\ldots,K$ and (ii) if $\beta_T \to 0$, but $\beta_T T \to \infty$, the variates are asymptotically independent normals with variances $2\beta_T T\, p_{a_k b_k}(u_k)$, $k = 1,\ldots,K$.

The two asymptotic distributions are consistent for large $\beta_T T$, because a Poisson variate with large mean is approximately normal. The result in (i) is not unexpected because we are counting "rare" events. It is surprising that such a general result is so simple however.

The theorem leads us to estimate $p_{ab}(u)$ by

$$\hat{p}_{ab}(u) = J_{ab}^T(u)/(2\beta_T T)$$

and to approximate the distribution of this variate by

$$(2\beta_T T)^{-1} P(2\beta_T T p_{ab}(u)) \text{ or } N(p_{ab}(u), (2\beta_T T)^{-1} p_{ab}(u)),$$

where $P(\mu)$ here denotes a Poisson distribution with

mean μ. This estimate should prove reasonable so long as $|u| \ll T$. In the case that u has noticeable magnitude compared to T it might be better to replace $J_{ab}^T(u)$ by $T\, J_{ab}^T(u)/(T-|u|)$ or by

$$J_{ab}^T(u) + p_a p_b |u| 2\beta_T \tag{8}$$

The use of the variate of (8) is suggested by the usual estimate of the autocovariance function of an ordinary time series. Its construction is based on the observation that $q_{ab}(u) \to 0$ as $|u| \to \infty$ for many processes. It should have better overall mean-squared error properties for such processes.

We remark that we are here essentially carrying out histogram construction. Considerations of that topic are relevant here. For example, we might choose to construct a rootogram based on $\sqrt{J_{ab}^T(u)}$ to get stable variance. (If there may be some cells with low counts, we might follow Tukey and use $\sqrt{2 + 4\, J_{ab}^T(u)}$). The variate $\sqrt{\hat{p}_{ab}(u)}$ will have approximately stable variance of $(8\beta_T T)^{-1}$. The theorem likewise leads us to estimate $p_{ab}(u)/p_b$ by $J_{ab}^T(u)/(2\beta_T N_b(T))$ and to approximate the distribution of this estimate by

$$(2\beta_T T\, p_b)^{-1} P(2\beta_T T\, p_{ab}(u)) \text{ or } N(p_{ab}(u)/p_b,$$

$$(2\beta_T T\, p_b^2)^{-1} p_{ab}(u)).$$

The variance of $\sqrt{J_{ab}^T(u)/(2\beta_T N_b(T))}$ will be approximately stable and may be estimated by $(8\beta_T N_b(T))^{-1}$.

The above results may be used to set

approximate confidence intervals and multiple confidence intervals for the estimates. In the case that the increment of the process N_a is independent of the increment of the process N_b, u time units away, $p_{ab}(u)/p_b = p_a$. We may examine this hypothesis by plotting

$$\sqrt{J_{ab}^T(u)/(2\beta_T N_b(T))} \;,\; \sqrt{\hat{p}_a} \;,\; \sqrt{J_{ab}^T(u)/(2\beta_T N_b(T))} \;\pm\; (2\beta_T N_b(T))^{-\frac{1}{2}}$$

on the same graph for example. This sort of graph is useful in checking for some degree of association between the process N_a and the process N_b.

What we have been doing may be viewed as estimating the probability density function of the times between a events and b events from the observed differences

$$\tau_j^a - \tau_k^b \;;\; 0 < \tau_j^a,\; \tau_k^b \leq T.$$

Cox (1965) suggested that one could also consider "window estimates." Let $W(u)$ be bounded and absolutely integrable. Let $W^T(u) = W(u/\beta_T)$ for the sequence of scale factors β_T, $T = 1, 2, \ldots$. It is now natural to base estimates on

$$J_{ab}^T(u) = \sum_{0 < \tau_j^a \neq \tau_k^b \leq T} W^T(u - \tau_j^a + \tau_k^b)$$

$$= \iint_{0 < \tau \neq \sigma \leq T} W^T(u - \sigma + \tau)\, dN_a(\sigma)\, dN_b(\tau)$$

(The previous $J_{ab}^T(u)$ corresponds to $W(u) = 1$ for

$|u| < 1$.) The variances of the asymptotic distributions of (ii) of Theorem 2 are now replaced by $\beta_T T \int W(u)^2 du\, p_{ab}(u_k)$, $k = 1,\ldots,K$. By direct computation we see that

$$E\, J^T_{ab}(u) = \int_{-T}^{T}(T - |\rho|)W^T(u - \rho)\, p_{ab}(\rho)d\rho$$

$$\sim \beta_T(T - |u|)\{p_{ab}(u)\int W(\rho)d\rho - \beta_T p'_{ab}(u)$$

$$\int \rho W(\rho)d\rho + \beta_T^2 p''_{ab}(u)\int \rho^2 W(\rho)d\rho/2 +..\}$$

suggesting that bias may become a problem when $p_{ab}(\rho)$ varies substantially in the neighborhood of u or when u is of appreciable magnitude compared to T. We have already discussed one modification to handle this last case.

The asymptotic distribution determined in Theorem 2 is an unconditional one. In practise the worker may feel that the conditional distribution, conditional on the observed $N_a(T)$, $N_b(T)$ is the appropriate one. In I.4 we set down the form of product densities in the conditional case. It should be possible to make use of these to determine the form of the large sample conditional distribution.

Cox and Lewis (1972) discuss some aspects of the problem of estimating second-order product densities for a vector-valued process.

3. In the k-th order case we might consider the statistic

$$J^T_{a_1\ldots a_k}(u_1,\ldots,u_{k-1}) = \int_0^T \cdots \int_0^T W^T(u_1-\sigma_1+\sigma_k,\ldots,$$

$$u_{k-1}-\sigma_{k-1}+\sigma_k) \, dN_{a_1}(\sigma_1)\ldots dN_{a_k}(\sigma_k) \quad (9)$$

where $W^T(u_1,\ldots,u_{k-1}) = W(u_1/\beta_T,\ldots,u_{k-1}/\beta_T)$ and the $\neq$ in (9) indicates that the range of integration is over distinct σ_j.

Theorem 3. Suppose the process $\underline{N}$ satisfies Assumption I and that $p_{a_1\ldots a_k}(\cdot)$ is continuous at $(u_1,\ldots,u_{k-1})$. Then as $T \to \infty$, (i) if $\beta_T^{k-1}T = L$, L fixed, if $W(u_1,\ldots,u_{k-1}) = 1$ for $|u_j| < 1$, the variate of (9) is asymptotically Poisson with mean $(2\beta_T)^{k-1}T \, p_{a_1\ldots a_k}(u_1,\ldots,u_{k-1})$ and (ii) if $\beta_T \to 0$, but $\beta_T^{k-1}T \to \infty$, the variate is asymptotically normal with mean

$$T\int_{-T}^T \cdots \int_{-T}^T W^T(u_1-\rho_1,\ldots,u_{k-1}-\rho_{k-1}) p_{a_1\ldots a_k}$$

$$(\rho_1,\ldots,\rho_{k-1}) \, d\rho_1\ldots d\rho_{k-1} \quad (10)$$

and variance $\beta_T^{k-1}T \int W^2 \, p_{a_1\ldots a_k}(u_1,\ldots,u_{k-1})$.

The integral of (10) may be expected to be near

$$\beta_T^{k-1} \, T^{k-1} \int W \, p_{a_1\ldots a_k}(u_1,\ldots,u_{k-1})$$

suggesting the consideration of the estimate

$$\hat{p}_{a_1\ldots a_k}(u_1,\ldots,u_{k-1}) = J^T_{a_1\ldots a_k}(u_1,\ldots,u_{k-1})/$$

$$(\beta_T^{k-1} \, T^{k-1} \int W)$$

4. Let A denote the interval $(0,T]$ and suppose that the points observed in A are: γ_1 of type 1 at $t_1,\ldots,t_{\gamma_1}$; $\ldots$; γ_r of type r at $\ldots, t_k$. Then, using the expressions of Section I, the likelihood function here is B/C where

$$B = \sum_{\nu_1 \geq \gamma_1} \cdots \sum_{\nu_r \geq \gamma_r} \frac{(-1)^{\nu_1-\gamma_1+\ldots+\nu_r-\gamma_r}}{(\nu_1-\gamma_1)!\cdots(\nu_r-\gamma_r)!}$$

$$\int_A^{\nu_1-\gamma_1+\ldots+\nu_r-\gamma_r} p^{(\nu_1)\ldots(\nu_r)}$$

$$(t_1,\ldots,t_{\gamma_1};\ldots:t_{\gamma_1+1},\ldots,t_{\gamma_1+\gamma_2};\ldots:\ldots$$

and

$$C = \frac{1}{\gamma_1!\cdots\gamma_r!} \sum_{\nu_1 \geq \gamma_1} \cdots \sum_{\nu_r \geq \gamma_r} \frac{(-1)^{\nu_1-\gamma_1+\ldots+\nu_r-\gamma_r}}{(\nu_1-\gamma_1)!\cdots(\nu_r-\gamma_r)!}$$

$$\int_A^{\nu_1+\ldots+\nu_r} p^{(\nu_1)\ldots(\nu_r)}$$

Let us consider the approximate value of the likelihood function, B/C, for large T. In the case of large T

$$\int_A^{\nu_1-\gamma_1+\ldots+\nu_r-\gamma_r} p^{(\nu_1)\ldots(\nu_r)} (t_1,\ldots t_{\gamma_1};\ldots$$

$$\sim p^{(\gamma_1)\ldots(\gamma_r)}(t_1,\ldots,t_{\gamma_1+\ldots+\gamma_r})$$

$$T^{\nu_1-\gamma_1+\ldots+\nu_r-\gamma_r} p_1^{\nu_1-\gamma_1} \cdots p_r^{\nu_r-\gamma_r}$$

suggesting that for large T, the likelihood function is approximately

$$\gamma_1! \ldots \gamma_r! \, p^{(\gamma_1)} \ldots ^{(\gamma_r)}(t_1,\ldots,t_{\gamma_1+\ldots+\gamma_r})$$
$$(Tp_1)^{-\gamma_1} \ldots (Tp_r)^{-\gamma_r}$$

4. In this section we will propose estimates of the parameters described in Section I.5 in the case that a realization of a stationary process is available for the time interval $(0,T]$.

(i) We begin with the **renewal** function,

$$U_{ab}(t) = E\{N_a(t) \mid N_b\{0\} = 1\} = \int_0^t p_{ab}(u) \, du / p_b$$

A natural estimate to consider is

$$\hat{U}_{ab}(t) = \#\{t \geq \tau_j^a - \tau_k^b > 0\}/N_b(T)$$
$$= \int_0^{T-t} \int_0^t dN_a(u+w) \, dN_b(u)/N_b(T)$$

To determine the asymptotic distribution of $\hat{U}_{ab}(t)$ we will need the joint asymptotic distribution of $\#\{\cdot\}$ and $N_b(T)$. It is fairly clear that under Assumption I, the variate is asymptotically normal with asymptotic variance that is $O(T^{-1})$. However the form of the asymptotic variance seems very messy. In practise one would probably have to estimate it by segmenting the data.

(ii) Let us next estimate the survivor function

$$1 - F(t) = \text{Prob}\{N(t) = 0 \mid N\{0\} = 1\}$$
$$= \text{Prob}\{\tau_{i+1} - \tau_i > t\}$$
$$= 1 - \text{Prob}\{\tau_{i+1} - \tau_i \leq t\}$$

This last suggests the estimate

$$\hat{F}(t) = \#\{\tau_{i+1} - \tau_i \leq t\,;\, i = 1,\ldots, N(T)-1\}/N(T)$$

This estimate is based on the interarrival times $X_i = \tau_{i+1} - \tau_i$. The process X_i, $i = 0, \pm 1,\ldots$ is stationary. If it is mixing in some sense then $1 - \hat{F}(t)$ will be asymptotically normal, see Deo (1973), for example. This last suggests the interesting problem of relating a mixing condition for a stationary point process to some mixing condition for the corresponding process of interarrival times.

(iii) The following is a plausible estimate for the hazard function, with β_T a small positive number,

$$\hat{\mu}(t) = \frac{\#\{t-\beta_T < \tau_{i+1} - \tau_i < t+\beta_T\,;\, i=1,\ldots,N(T)-1\}}{2\beta_T \#\{\tau_{i+1} - \tau_i > t\,;\, i=1,\ldots,N(T)-1\}}$$

(iv) Next consider the estimation of the forward recurrence time distribution

$$\begin{aligned}G(t) &= 1 - \mathrm{Prob}\{N(t) = 0\} \\ &= p\int_0^t (1 - F(u))\,du \\ &= p[(1 - F(t))t] + p\int_0^t u\,dF(u)\end{aligned}$$

where we use a Palm-Khinchin relation from Section I.9 and integrate by parts. The last relation suggests the estimate

$$\hat{G}(t) = \hat{p}[(1-\hat{F}(t))t] + \hat{p} \sum_{\tau_{i+1}-\tau_i \leq t} (\tau_{i+1}-\tau_i)/(N(T)-1)$$

$$\sim t \, \#\{\tau_{i+1}-\tau_i > t\}/T + \sum_{\tau_{i+1}-\tau_i \leq t} (\tau_{i+1}-\tau_i)/T$$

III. Estimation of Frequency Domain Parameters

1. We begin with a discussion of first order statistics. Suppose $T = JU$, J an integer. Set

$$d_a^U(\lambda,j) = \int_{jU}^{(j+1)U} \exp\{-i\lambda t\} \, dN_a(t) \qquad j = 0,\ldots, J-1$$

$$= \sum_{jU < \tau_a \leq (j+1)U} \exp\{-i\lambda \tau_a\}$$

$$= \int \exp\{-i(\lambda-\alpha)(j+\tfrac{1}{2})\}(\sin(\lambda-\alpha)U/2)/((\lambda-\alpha)/2) \, dZ_a(\alpha)$$

using the spectral representation at the last step. In the case that $J = 1$, $U = T$, we shall write $d_a^T(\lambda)$. We have,

Theorem 4. Let the process $\underset{\sim}{N}(t)$ satisfy Assumption I. Suppose $\lambda \neq 0$. Then $\underset{\sim}{d}^U(\lambda,j)$, $j = 0,\ldots,J-1$ are asymptotically independent r variate complex normal with mean $\underset{\sim}{0}$ and covariance matrix $2\pi U[f_{ab}(\lambda)]$ as $T \to \infty$. Also variates at frequencies of the form $2\pi u/U$, are asymptotically independent for u distinct positive integers.

2. Suppose we are interested in estimating the second order spectrum $f_{ab}(\lambda)$. Various procedures

suggest themselves, based either on the expression

$$E\{dZ_a(\lambda) \, dZ_b(\mu)\} = \delta(\lambda-\mu) \, f_{ab}(\lambda) \, d\lambda \, d\mu \quad \lambda,\mu \neq 0$$

or the expression

$$\{p_a + \int_{-\infty}^{\infty} \{p_{aa}(u) - p_a^2\} \exp\{-i\lambda u\} \, du\}/(2\pi) = f_{aa}(\lambda)$$

$$\{\int_{-\infty}^{\infty} \{p_{ab}(u) - p_a p_b\} \exp\{-i\lambda u\} \, du\}/(2\pi) = f_{ab}(\lambda) \quad a \neq b$$

<u>Procedure I</u>. Set $\underline{I}^U(\lambda,j) = (2\pi U)^{-1} \underline{d}^U(\lambda,j) \overline{\underline{d}^U(\lambda,j)}^T$ for $\lambda \neq 0$ and consider the estimate

$$\underline{f}^U(\lambda) = J^{-1} \sum_{j=0}^{J-1} \underline{I}^U(\lambda,j)$$

From Theorem 4, as $T \to \infty$, but J remains fixed $\underline{f}^U(\lambda)$ tends to $J^{-1} W_r^C(J, \underline{f}(\lambda))$ where W_r^C denotes the complex Wishart.

<u>Procedure II</u>. Set $\underline{I}^T(\lambda) = (2\pi T)^{-1} \underline{d}^T(\lambda) \overline{\underline{d}^T(\lambda)}^T$. For $2\pi s_j/T$ distinct, $\neq 0$, non-negative and all $\to \lambda$ set

$$\underline{f}^T(\lambda) = J^{-1} \sum_{j=0}^{J-1} \underline{I}^T(2\pi s_j/T).$$

From Theorem 4, as $T \to \infty$, $\underline{f}^T(\lambda)$ tends to $J^{-1} W_r^C(J, \underline{f}(\lambda))$

Both of the above estimates are asymptotically normal if the limiting conditions are as $T \to \infty$, $J \to \infty$, but $J/T \to 0$.

In the above procedures we sometimes choose to weight the periodogram ordinates unequally. For example in Procedure II we might take

$$\underline{f}^T(\lambda) = \frac{2\pi}{T} \sum_{s \neq 0} W^T(\lambda - \frac{2\pi s}{T}) \underline{I}^T(\frac{2\pi s}{T})$$

with $W^T(\alpha) = B_T^{-1} W(\alpha/B_T)$ where $\int W = 1$. If $B_T \to 0$, $B_T T \to \infty$ as $T \to \infty$, this estimate is asymptotically normal, see Brillinger (1972).

<u>Procedure III.</u> Let $\hat{p}_a$, $\hat{p}_{ab}(u)$ be given by the expressions of II.1, II.2 respectively. Let $w^T(u) = w(B_T u)$ be a convergence factor. Set

$$f_{ab}^T(\lambda) = \{2\beta_T \sum_j \{\hat{p}_{ab}(2\beta_T j) - \hat{p}_a \hat{p}_b\} \exp\{-i\lambda 2\beta_T j\}$$

$$w^T(2\beta_T j)\}/(2\pi) \qquad a \neq b$$

$$= \{\hat{p}_a + 2\beta_T \sum_j \{\hat{p}_{aa}(2\beta_T j) - \hat{p}_a^2\} \exp\{-i\lambda 2\beta_T j\}$$

$$w^T(2\beta_T j)\}/(2\pi) \qquad a = b$$

Because of the periodicities involved, it only makes sense to compute this estimate for $|\lambda| \leq \pi/\beta_T$. The choice of bin width $2\beta_T$ is seen to show itself in the Nyquist frequency π/β_T. This estimate is asymptotically normal under conditions including B_T, $\beta_T \to 0$, $B_T T \to \infty$ as $T \to \infty$. This estimate is the one computed most rapidly. It has the disadvantage of possibly leading to negative power spectrum estimates and coherences bigger than 1, even if $W(\alpha) \geq 0$.

<u>Procedure IV.</u> Compute the spectrum of the ordinary process

$$X(t) = h^{-1} N(t - \frac{h}{2}, t + \frac{h}{2}) \qquad t = 0, \pm h, \pm 2h, \ldots$$

but remember that

$$f_{XX}(\lambda) = \sum_j f_{NN}(\lambda + \tfrac{2\pi j}{h})(\sin(\lambda + \tfrac{2\pi j}{h}))^2 / (\lambda + \tfrac{2\pi j}{h})^2$$

Problems of aliasing clearly arise here.

Tapering and prefiltering play essential roles in the estimation of the spectra of ordinary time series. It is not entirely obvious how to apply these techniques in the point process case (with the exception of tapering for Procedures I and II.)

If the complete intensity

$$\lambda(t)h \sim \text{Prob}\{\text{point in }(t,t+h)\} \mid N(u), u \le t\}$$

exists and can be evaluated, then with

$$\Lambda(t) = \int_0^t \lambda(t)\,dt$$

the transformation $N(t) \rightarrow N(\Lambda(t))$ carries N over into a Poisson process with unit intensity, and constant power spectrum. (This transformation is analagous to the conditional probability integral transformation to uniform variates in the case of ordinary time series.) For the doubly stochastic Poisson process $\lambda(t) = x(t)$.

Prefiltering procedures carried out entirely in the frequency domain, for ordinary time series, clearly have point process analogs. For example, if we can think of a $g(\lambda)$ near $f(\lambda)$, then we might form the estimate

$$g(\lambda)\,\tfrac{2\pi}{T} \sum_{s \ne 0} W^T(\lambda - \tfrac{2\pi s}{T})\, g(\tfrac{2\pi s}{T})^{-1}\, I^T(\tfrac{2\pi s}{T})$$

Detrending can be very important. Lewis (1972) contains important advice on these matters.

3. We next turn to a brief discussion of the estimation of the parameters of the model

$$\text{Prob}\{\text{type 1 event in } (t,t+h] \mid N_2(u), u \leq t\}$$
$$\sim (\mu + \int a(t-u) \, dN_2(u))h$$

as $h \downarrow 0$. If $p_{22}(u)$ is not constant, then we estimate $a(u)$, a time domain parameter by going through the frequency domain. We have the relations

$$A(\lambda) = \int a(u) \exp\{-i\lambda u\} du$$
$$p_1 = \mu + A(0)p_2$$
$$f_{12}(\lambda) = A(\lambda) f_{22}(\lambda)$$
$$a(u) = (2\pi)^{-1} \int A(\alpha) \exp\{iu\alpha\} d\alpha$$

suggesting the estimates

$$\hat{A}(\lambda) = f_{12}^T(\lambda)/f_{22}^T(\lambda)$$
$$\hat{\mu} = \hat{p}_1 - \hat{A}(0) \hat{p}_2$$
$$\hat{a}(u) = (2\pi)^{-1} B_T \sum_k \hat{A}(kB_T) \exp\{iukB_T\} v^T(kB_T)$$

where $v^T(\alpha) = v(C_T \alpha)$ is a convergence factor. More details on this procedure may be found in Brillinger (1974 a).

4. On occasion we may be led to model the process in some manner involving a finite dimensional parameter θ. We would then like to be able to estimate θ. Sometimes such a model will lead to a tractable form

for the second-order spectra. For example, suppose we have a cluster process with primary process Poisson and the secondary process independent exponentials from the cluster centers, then the power spectrum of the process has the form

$$f(\lambda; \theta) = a^2 \frac{(b^2 + \lambda^2)}{(c^2 + \lambda^2)}$$

involving the three dimensional parameter $\theta = (a,b,c)$. We now describe one method of estimating θ. Related methods are given in Whittle (1953), Walker (1964), Hawkes and Adamopoulos (1973).

Let the true value of θ be θ'. Suppose

$$\lim_{|\lambda| \to \infty} f(\lambda; \theta) = \mu(\theta)$$

and $\mu(\theta') = p/(2\pi)$ where p is the mean intensity of the process. $\mu(\theta')$ may be estimated by $\hat{\mu} = \hat{p}/(2\pi)$.

The periodograms $I^T(2\pi s/T)$, $s = 1,2,\ldots$ are asymptotically independent exponentials with means $f(2\pi s/T; \theta')$, $s = 1,2,\ldots$. The scaled variates

$$I^T(2\pi s/T)/\hat{\mu} \qquad s = 1,2,\ldots$$

are therefore asymptotically independent exponentials with means

$$f(2\pi s/T; \theta')/\mu(\theta') \qquad s = 1,2,\ldots$$

This result suggests our setting down the following approximate "log likelihood" function

$$-\sum_{s=1}^{S_T}\left\{\log\frac{f(2\pi s/T;\theta)}{\mu(\theta)}+\frac{\mu(\theta)}{f(2\pi s/T;\theta)}\frac{I^T(2\pi s/T)}{\hat{\mu}}\right\} \quad (11)$$

and taking as an estimate of θ, the value $\hat{\theta}$ that maximizes (11).

In the theorem below we set $g(\lambda;\theta) = f(\lambda;\theta)/\mu(\theta)$ and

$$\Lambda^T(\theta) = -\frac{2\pi}{T}\sum_{s=1}^{S_T}\left\{\log g(2\pi s/T;\theta) + \frac{I^T(2\pi s/T)/\hat{\mu}}{g(2\pi s/T;\theta)} - 1\right\} \quad (12)$$

The $\hat{\theta}$ maximizing (11) also maximizes (12).

<u>Theorem 5.</u> If (a) the process $N(t)$, $-\infty < t < \infty$, has mean intensity $\mu(\theta')$ and power spectrum $f(\lambda;\theta')$, (b) $f(\lambda;\theta)$, $\theta \in \Theta \subset R^L$, is non-negative and

$$\mu(\theta) = \lim_{|\lambda|\to\infty} f(\lambda;\theta)$$

exists, (c) with $g(\lambda;\theta) = f(\lambda;\theta)/\mu(\theta)$,

$$\Lambda(\theta) = -\int\{\log g(\lambda;\theta) + \frac{g(\lambda;\theta')}{g(\lambda;\theta)} - 1\}\,d\lambda$$

exists as a Lebesgue integral, has a unique maximum at θ' and is such that

$$\max_{\theta''\in U} \Lambda(\theta'') \to \Lambda(\theta)$$

as the neighborhood U of θ shrinks to $\{\theta\}$, (d) $\Lambda^T(\theta) \xrightarrow{P} \Lambda(\theta)$ at θ' and uniformly near other θ (e) $\theta\in\Theta$ maximizing (11) is bounded in probability, then $\hat{\theta} \to \theta'$ in probability as $T \to \infty$.

Condition (d) is satisfied for processes satisfying Assumption I, provided $g(\lambda;\theta)$ is a sufficiently regular function of λ.

We next turn to the large sample distribution of $\hat{\theta}$. To this end set,

$$\Lambda_j^T(\theta) = \frac{\partial}{\partial \theta_j} \Lambda^T(\theta) \qquad \Lambda_{jk}^T(\theta) = \frac{\partial^2}{\partial \theta_j \partial \theta_k} \Lambda^T(\theta) \qquad (13)$$

Because of (c) above, generally $\Lambda_j(\theta') = \partial \Lambda(\theta')/\partial \theta_j = 0$.

<u>Theorem 6.</u> Suppose the conditions of Theorem 5 are satisfied. Suppose also (f) the derivatives of (13) exist, (g) $\Lambda_{jk}^T(\zeta^T) \to A_{jk}$ for any sequence ζ^T of variates tending to θ' in probability, (h) with $\underset{\sim}{A} = [A_{jk}]$, $\sqrt{T}\{\Lambda_1^T(\theta'),\ldots,\Lambda_L^T(\theta')\} \to N_L(\underset{\sim}{0}, \underset{\sim}{A}+\underset{\sim}{B})$, then $\hat{\theta}$ is asymptotically normal with mean θ' and covariance matrix $T^{-1}\underset{\sim}{A}^{-1}(\underset{\sim}{A}+\underset{\sim}{B})\underset{\sim}{A}^{-1}$.

For processes satisfying Assumption I and $g(\lambda;\theta)$ a sufficiently regular function of λ we have

$$A_{jk} = \int_0^\infty \frac{\partial \log g(\alpha;\theta')}{\partial \theta_j'} \frac{\partial \log g(\alpha;\theta')}{\partial \theta_k'} d\alpha$$

$$B_{jk} = 2\pi \int_0^\infty \int_0^\infty \frac{\partial \log g(\alpha;\theta')}{\partial \theta_j'} \frac{\partial \log g(\beta;\theta')}{\partial \theta_k'} \frac{f^{(4)}(\alpha,-\alpha,-\beta;\theta')}{f(\alpha;\theta') f(\beta;\theta')} d\alpha \, d\beta$$

The above procedure provides us with a further estimate, $\hat{f}(\lambda) = f(\lambda;\hat{\theta})$ of the power spectrum. Under the conditions of the theorem, this estimate will be asymptotically normal with mean $f(\lambda;\theta')$ and variance

$$\sum_{j,k} \frac{\partial f(\lambda;\theta')}{\partial \theta'_j} \frac{\partial f(\lambda;\theta')}{\partial \theta'_k} \text{cov}\{\hat{\theta}_j, \hat{\theta}_k\}$$

In the case of a vector-valued process, instead of maximizing (11) we would maximize

$$-\sum_{s=1}^{S} \{\log \text{Det } \underline{g}(2\pi s/T;\theta) + \text{tr}(\underline{H}^T(2\pi s/T) \underline{g}(2\pi s/T;\theta))\}$$

where

$$g(\lambda;\theta)_{jk} = f(\lambda;\theta)_{jk}/\sqrt{\mu_j(\theta)\mu_k(\theta)}$$

$$H^T(\lambda)_{jk} = I^T(\lambda)_{jk}/\sqrt{\hat{u}_j \hat{u}_k}$$

5. We mention briefly that the parameters of a self-exciting process may be estimated via a frequency domain analysis. Such a process is defined by a relationship

$$E\{dN(t) \mid N(u), u \le t\} = (\mu + \int_{-\infty}^{t} a(t-u)dN(u)) \, dt$$

where μ, $a(u) \ge 0$; $\int a(u) \, du < 1$; $a(u) = 0$ for $u < 0$. Let

$$A(\lambda) = \int_0^\infty a(u) \exp\{-i\lambda u\} du$$

For this process $\mu = p[1 - A(0)]$ and

$$f(\lambda) = p/(2\pi|1 - A(\lambda)|^2)$$

Because $A(\lambda)$ is the Fourier transform of a one-sided function, the problem of estimating $A(\lambda)$ from $f^T(\lambda)$, is seen to involve the factorization of $f^T(\lambda)$. Rice (1973) carried out this empirically and found the asymptotic distribution of the estimate. This procedure also provides a further spectral estimate, namely $\hat{f}(\lambda) = \hat{p}/(2\pi|1-\hat{A}(\lambda)|^2)$.

6. We next turn to the problem of estimating the variance time curve given by var $N(t)$ as a function of t. Using the spectral representation, we see that

$$V(t) = \text{var } N(t) = \int_{-\infty}^{\infty} \left(\frac{\sin \alpha t/2}{\alpha/2}\right)^2 f(\alpha) \, d\alpha$$

$$= tp + \int_{-\infty}^{\infty} \left(\frac{\sin \alpha t/2}{\alpha/2}\right)^2 \left(f(\alpha) - \frac{p}{2\pi}\right) d\alpha$$

The following type of estimate is considered by Torres-Melo (1974),

$$V(t) = t\hat{p} + B\left[t^2\left(f^T(0) - \frac{\hat{p}}{2\pi}\right) + 2\sum_{s=1}^{S}\left(\frac{\sin Bst/2}{Bs/2}\right)^2 \left(f^T(Bs) - \frac{\hat{p}}{2\pi}\right)\right]$$

He finds the asymptotic distribution of this estimate.

7. Product densities may be estimated in a similar manner to the variance time curve. We have

$$p(u) = \int_{-\infty}^{\infty} (f(\alpha) - \frac{p}{2\pi}) \exp\{iu\alpha\} \, d\alpha + p^2$$

suggesting the estimate

$$\hat{p}(u) = B[(f^T(0) - \frac{\hat{p}}{2\pi}) + 2\sum_{s=1}^{S}(f^T(Bs) - \frac{\hat{p}}{2\pi})\cos Bs]$$

This estimate would undoubtedly be improved by the insertion of convergence factors.

Finally we remark that we may sometimes wish to estimate the spectral measure

$$F(\lambda) = \int_0^\lambda f(\alpha)\, d\alpha$$

The obvious estimate is

$$\hat{F}(\lambda) = B \sum_{Bs \le \lambda} f^T(Bs)$$

IV. Proofs

1. <u>Proof of Theorem 1.</u> The joint factorial cumulant of $N_{a_1}(T),\ldots,N_{a_k}(T)$ is

$$\int_0^T \cdots \int_0^T q_{a_1 \cdots a_k}(t_1,\ldots,t_k)\, dt_1 \cdots dt_k = O(T)$$

in view of Assumption I. The ordinary joint cumulant of these same variates is a sum of multiples of lower order factorial cumulants. It follows that it too is $O(T)$ as $T \to \infty$. This means that the standardized joint cumulants of order k of these variates are $O(T^{1-k/2}) \to 0$ as $T \to \infty$ for $k > 2$, and so the variates are asymptotically jointly normal.

2. <u>Proof of Theorem 2.</u> The variate $J^T_{ab}(u)$ may be represented as

$$\int_G dN_a(\sigma)\, dN_b(\tau)$$

where G is the set $\{u - \beta_t < \sigma - \tau < u + \beta_T, \sigma \ne \tau\}$. It follows from this representation, Assumption I and the rules of Leonov and Shiryaev (1959) that the joint factorial moment of order k of $J_{ab}^T(u)$ is of order $O(\beta_T^k T)$. An ordinary cumulant of order k, c_k, is connected to corresponding factorial cumulants, $c_{(k)}$, through

$$c_k = \sum_{j=1}^{k} S_j^k \, c_{(j)}$$

where S_j^k is a Stirling number. If $\beta_T = L/T$, then $E\, J_{ab}^T(u^T) \to 2L\, p_{ab}(u)$ as $T \to \infty$, when $u^T \to u$. It follows, that in this case the cumulant of order k of $J_{ab}^T(u^T) \to 2L\, p_{ab}(u)$ and so the variate is asymptotically Poisson. In the case $\beta_T T \to \infty$, the standardized joint cumulant of order k is $O(\beta_T T)^{1-k/2} \to 0$ for $k > 2$. It follows that the variate is asymptotically normal. The indicated asymptotic independence follows on evaluating joint second-order cumulants.

3. Theorem 3 is proved in the same manner that Theorem 2 is proved.

4. Theorem 4 is proved by evaluating the joint cumulants of the d_a^U. A related result, Theorem 4.2, is proved in Brillinger (1972).

5. Before proving Theorem 5, we prove a lemma of some independent interest.

Lemma 1. If (i) Θ is locally compact, complete, separable, metric, (ii) $(\Omega, \mathcal{G}, P)$ is a probability space with Ω complete, separable, metric, (iii) $Q^T(\theta, \omega)$ is real-valued, Borel measurable for $(\theta, \omega) \in \Theta \times \Omega$ and all T, (iv) $Q(\theta)$ is real-valued, lower semi-continuous, $Q(\theta) > Q(\theta')$ for $\theta \neq \theta'$, (v)

$Q^T(\theta',w) = Q(\theta') + o_p(1)$, $Q^T(\theta,w) \geq Q(\theta) + o_p(1)$, $\theta \neq \theta'$ as $T \to \infty$, (vi) given $\varepsilon, \eta > 0$, $\theta_1 \neq \theta'$, there exists U_1 a neighborhood of θ_1 and there exists T_0 such that

$$\text{Prob}\{[Q^T(\theta_1,w) - \inf_{\theta \in U_1} Q^T(\theta,w)] \leq -\varepsilon\} < \eta \text{ for } T \geq T_0$$

(vii) for each w and T there exists $\hat{\theta}$ such that

$$Q^T(\hat{\theta},w) = \inf_{\theta \in \Theta} Q^T(\theta,w)$$

(viii) given $\eta > 0$, there exists a compact set $C \subseteq \Theta$ and T_0 such that $\text{Prob}\{\hat{\theta} \notin C\} < \eta$, for $T > T_0$, then $\hat{\theta} = \theta' + o_p(1)$.

Proof. The measurability of $\hat{\theta}$ results from Theorem 2 of Brown and Purves (1973). Let $U \subseteq C$ be an open neighborhood of θ'. From (iv) there exists $\gamma > 0$ such that $Q(\theta_1) - Q(\theta') \geq 3\gamma$ for $\theta \in C \setminus U$. Suppose $\theta_1 \in C \setminus U$. Then from (v)

$$\lim_{T \to \infty} \text{Prob}\{Q^T(\theta_1,w) - Q^T(\theta',w) \leq 2\gamma\} = 0$$

From this and (v) there exists a neighborhood U_1 of θ_1 such that

$$\lim_{T \to \infty} \text{Prob}\{\inf_{\theta \in U_1} Q^T(\theta,w) - Q^T(\theta',w) \leq \gamma\} = 0 \tag{14}$$

Using the fact that C is compact, select a finite number of points θ_s, $s = 1,\ldots,N$ with neighborhoods U_s, $s = 1,\ldots,N$ covering $C \setminus U$. From (14)

$$\lim_{T \to \infty} \text{Prob}\{ \inf_{\theta \in C\setminus U} Q^T(\theta,\omega) - Q^T(\theta',\omega) \le \gamma \} = 0 \quad (15)$$

Now from (vii)

$$\text{Prob}\{\hat{\theta} \notin U \text{ or } \Theta\setminus C\} \le \text{Prob}\{ \inf_{\theta \in C\setminus U} Q^T(\theta,\omega) - Q^T(\theta',\omega) \le \gamma \}$$

From (15) this last tends to 0. From (viii), Prob$\{\hat{\theta} \in \Theta\setminus C\}$ tends to 0. This gives the result.

Theorem 5 now follows from this Lemma.

6. Theorem 6 follows from the relation

$$\Lambda_j^T(\theta) = 0 = \Lambda_j^T(\theta') + \sum_k (\theta_k - \theta_k') \Lambda_{jk}^T(\alpha)$$

with α between θ and θ'.

REFERENCES

BRILLINGER, D. R. (1972). The spectral analysis of stationary interval functions. pp. 483-513 in Proc. Sixth Berkeley Symp. Math. Stat. Prob., Vol. I. (eds. L. M. LeCam, J. Neyman, E. L. Scott). Berkeley, University of California Press.

BRILLINGER, D. R. (1974a). Cross-spectral analysis of processes with stationary increments including G/G/∞ queue. Ann. Prob., 2, 815-827.

BRILLINGER, D. R. (1974b). The identification of point process systems. Special Invited Lecture presented to the Institute of Mathematical Statistics at Edmonton.

COX, D. R. (1965). On the estimation of the intensity function of a stationary point process. J. R. Statist. Soc., B, 27. 332-337.

COX, D. R. and LEWIS, P. A. W. (1966). The Statistical Analysis of Series of Events. London, Methuen.

COX, D. R. and LEWIS, P. A. W. (1972). Multivariate point processes. pp. 401-448 in Proc. Sixth Berkeley Symp. Math. Stat. Prob., Vol. II. (eds. L. M. LeCam, J. Neyman, E. L. Scott). Berkeley, University of California Press.

CRAMER, H., LEADBETTER, M. R. and SERFLING, R. J. (1971). On distribution function - moment relationships in stationary point processes. Zeit. Wahrschein., 18. 1-8.

DALEY, D. J. and VERE-JONES, D. (1972). A summary of the theory of point processes. pp. 299-383 in STOCHASTIC POINT PROCESSES (ed. P. A. W. Lewis)

New York, Wiley.

DEO, C. M. (1973). A note on empirical processes of strong-mixing sequences. Ann. Prob., 1. 870-875.

HAWKES, A. G. (1972). Spectra of some mutually exciting point processes with associated variables. pp. 261-271 in Stochastic Point Processes (ed. P. A. W. Lewis). New York, Wiley.

HAWKES, A. G. and ADAMOPOULOS, L. (1973). Cluster models for earthquakes - regional comparisons. Bul. Inter. Statist. Inst., 39.

LEONOV, V. P. and SHIRYAEV, A. N. (1959). On a method of calculation of semi-invariants. Theory Prob. Appl., 4. 319-329.

LEWIS, P. A. W. (1972). Recent results in the statistical analysis of univariate point processes. pp. 1-54 in Stochastic Point Processes (ed. P. A. W. Lewis). New York, Wiley.

RICE, J. A. (1973). Statistical analysis of self-exciting point processes and related linear models. Ph.D. Thesis, University of California, Berkeley.

SRINIVASAN, S. K. (1974). Stochastic Point Processes. New York, Hafner.

TORRES-MELO, L. (1974). Stationary point processes. Ph.D. Thesis, University of California, Berkeley.

WALKER, A. M. (1964). Asymptotic properties of least squares estimates of parameters of the spectrum of a stationary nondeterministic time series. J. Austral. Math. Soc., 4. 363-384.

WHITTLE, P. (1953). Estimation and information in stationary time series. Ark. Math. Astron. Fys., 2. 423-434.

ASPECTS OF EXTREME VALUE THEORY
FOR STATIONARY PROCESSES - A SURVEY

by M. R. Leadbetter
Department of Statistics
University of North Carolina at Chapel Hill

SUMMARY

The primary concern in this paper is with the distributional results of classical extreme value theory, and their development to apply to stationary processes. The main emphasis is on stationary sequences, where the theory is well developed. Results available for continuous parameter processes are also described, but with particular reference to stationary normal processes.

Research sponsored by Office of Naval Research under Contract N00014-67-A-0321-0002 TSK NR-042-214.

1. INTRODUCTION

This paper is concerned with certain aspects of extreme value theory of independent and identically distributed (i.i.d.) random variables, and their generalization to apply to stationary sequences, and continuous parameter processes. If $\{\xi_n\}$ is any sequence of random variables we shall write

$$M_n = \max(\xi_1, \xi_2, \ldots, \xi_n)$$

for the maximum of the first n terms. Specifically the following topics from classical extreme value theory will be considered:

(i) Gnedenko's Theorem - giving the possible types of asymptotic distribution of M_n, i.e. the possible distributions G such that $P\{a_n(M_n - b_n) \leq x\} \to G(x)$

(ii) The relation $P\{M_n \leq u_n\} \to e^{-\tau}$ when $P(\xi_i > u_n) \sim \tau/n$

(iii) The form of the limiting distribution in the normal case

(iv) The Poisson limiting distribution for the number of exceedances of u_n by $\xi_1 \ldots \xi_n$ where u_n is chosen as in (ii)

(v) The asymptotic distribution of the rth largest of $\xi_1 \ldots \xi_n$ for fixed r.

In Section 2 these results will be stated (together with indications of methods of proof) for i.i.d. random variables. Section 3 contains a

brief discussion of dependence conditions which
have been used in the past, and those which are used
here, in order that the asymptotic results for the
maximum may apply to stationary sequences, as will
be seen in Section 4. These results will be
compared with those of Berman [1] for stationary
normal sequences.

In Section 5, we look at the "exceedances" of
high levels by stationary sequences, and indicate
their weak convergence as point processes, to a
Poisson process, by means of a simple and useful
general point process convergence theorem of
Kallenberg [13]. The desired properties of the rth
largest value, $M_n^{(r)}$, will follow as immediate
consequences of this convergence.

Certain of the results are known to apply to
the supremum $M(T) = \sup\{\xi(t): 0 \leq t \leq T\}$ of a
continuous parameter stationary process $\xi(t)$ as
$T \to \infty$. These are described in Section 6. The
most detailed properties presently known concern
the case when $\{\xi(t)\}$ is a normal process, and
these are also discussed in Section 6.

Finally we remark that other questions besides
distributional ones, are of interest in extreme
value theory. For example convergence questions
akin to laws of large numbers are clearly
important. We do not consider such matters here,
but concentrate on the distributional results
indicated.

2. EXTREME VALUE THEORY FOR I.I.D. RANDOM VARIABLES

In this section we shall look briefly at each
of the topics mentioned in the Introduction, for

i.i.d. random variables.

(i) Gnedenko's Theorem

This theorem is central to classical extreme value theory and is stated precisely as follows:

THEOREM 2.1 (GNEDENKO): Let $\xi_1, \xi_2 \ldots$ be i.i.d. random variables, and let $M_n = \max(\xi_1, \xi_2 \ldots \xi_n)$. Suppose that for some sequences of constants $\{a_n > 0\}$, $\{b_n\}$,

$$P\{a_n(M_n - b_n) \leq x\} \to G(x) \text{ (at continuity points of } G)$$

where G is a non-degenerate d.f.. Then G belongs to one of the three "extreme value types":

Type I $G(x) = \exp(-e^{-x})$ $-\infty < x < \infty$
Type II $G(x) = 0$ $x \leq 0$
         $= \exp(-x^{-\alpha})$ $x \geq 0$ $(\alpha > 0)$
Type III $G(x) = \exp(-(-x)^{\alpha})$ $x < 0$ $(\alpha > 0)$
          $= 1$ $x \geq 0$

(In saying that G is one of these "types" it is to be understood that x may be replaced by ax + b for any fixed a > 0, b in the functional form.)

While we shall not prove Gnedenko's Theorem here (see [10] or [5] for detailed proof) it does seem worthwhile to point out the main ideas of the proof, since these are also used later. The main part of the proof given by Gnedenko may be displayed as the three lemmas given below.

First note that if the common distribution function (d.f.) of the $\{\xi_i\}$ is F, then M_n has the d.f. $P\{\xi_1 \le x, \xi_2 \le x \ldots \xi_n \le x\} = F^n(x)$ and

$$P\{a_n(M_n - b_n) \le x\} = F^n\left(\frac{x}{a_n} + b_n\right).$$

If the left-hand side converges to some $G(x)$ then for any $k = 1,2,3\ldots$,

$$P\{a_{nk}(M_n - b_{nk}) \le x\} = F^n\left(\frac{x}{a_{nk}} + b_{nk}\right) \to G^{1/k}(x).$$

Stating this formally we have

LEMMA 2.2 Let $\xi_1, \xi_2 \ldots$ be i.i.d.. If

(2.1) $P\{a_{nk}(M_n - b_{bk}) \le x\} \to G^{1/k}(x)$ as $n \to \infty$

holds for $k = 1$, then it holds for all $k = 1,2,3\ldots$

In particular if the assumptions of Theorem 2.1 are satisfied then (2.1) holds for all k. On the other hand if (2.1) holds for all k, G must have certain "stability properties" indicated in the following lemma.

LEMMA 2.3 If (2.1) holds for all k and G is non-degenerate then for each k, there exist constants $\alpha_k > 0$, β_k such that

(2.2) $$G^k(\alpha_k x + \beta_k) = G(x) \text{ for all } x$$

(Sometimes such a d.f. is called stable (e.g. [5]) though this is, of course, not the usual definition of stability.)

This result is due to Khintchine and is given in [11]. A very general result of this type (from which this lemma follows as a special case) is given by Feller [7, p. 246]. This result states that if $\{F_n\}$ is any sequence of d.f.'s such that

$$F_n(\lambda_n x + \mu_n) \to G(x), \quad F_n(\lambda_n^* x + \mu_n^*) \to G^*(x),$$

for some sequences $\lambda_n > 0$, μ_n, $\lambda_n^* > 0$, u_n^*, where G, G^* are assumed non-degenerate, then
$\lambda_n/\lambda_n^* \to 1$, $(\mu_n - \mu_n^*)/\lambda_n \to 0$ and $G^*(x) = G(ax + b)$
for some $a > 0$, b. This theorem may be applied to (2.1) with $\lambda_n = 1/a_n$, $\mu_n = b_n$, $\lambda_n^* = 1/a_{nk}$, $\mu_n^* = 1/b_{nk}$, $G^* = G^{1/k}$, to give the lemma.

The following lemma now completes the proof of Gnedenko's Theorem for i.i.d. random variables.

<u>LEMMA 2.4</u> If G is a non-degenerate d.f. satisfying (2.2) for each k (and some constants $\alpha_k > 0$, β_k) then it is an extreme value d.f. of Type I, II, or III.

The proof of this lemma is the major part of Gnedenko's derivation in [10] and we do not give it here.

To summarize, if the ξ_n are i.i.d. it follows that

(a) If (2.1) holds for k = 1 it holds for all k = 1,2... . In particular (2.1) holds for all k under the assumptions of Gnedenko's Theorem.

(b) If (2.1) holds for all k then G (if non-degenerate) has the stability properties (2.2),

(c) If (2.2) holds for each k, G is a Type I, II, or III extreme value d.f. and hence

(d) under the assumptions of Gnedenko's Theorem, (2.1) holds and hence (2.2) does and hence G is an extreme value d.f. .

Looking ahead, we note that the same proof will apply under any circumstances in which the truth of (2.1) for $k = 1$ implies its truth for all k. This will be used in the next section to obtain the desired generalization of Gnedenko's Theorem.

(ii) Convergence of $P\{M_n \le u_n\}$ to $e^{-\tau}$ when $1 - F(u_n) \sim \tau/n$.

Suppose that a sequence $\{u_n\}$ may be chosen so that

(2.3) $\quad 1 - F(u_n) \ (= P\{\xi_1 > u_n\}) \sim \tau/n$

for some fixed $\tau > 0$. (Note that this is not always possible - **e.g.** if F increases only by jumps at 1,2,3,... with $F(j) = 1 - \tau/(2^j - 1)$, as may be checked.) Then

(2.4) $\quad P\{M_n \le u_n\} = [1 - (1 - F(u_n))]^n = [1 - \tau/n + o(1/n)]^n$
$\to e^{-\tau}$.

Conversely it is easily seen that (2.3) is also necessary for (2.4). (If (2.4) holds then

$$n \log[1 - (1 - F(u_n))] \to -\tau$$

from which (2.3) follows.) Thus the following result holds.

THEOREM 2.5. For the i.i.d. case $P\{M_n \leq u_n\} \to e^{-\tau}$

($\tau > 0$) if and only if $1 - F(u_n) \sim \tau/n$.

(iii) The limiting distribution in the normal case.

If ξ_i are i.i.d. standard normal r.v.'s we may choose u_n so that $1 - \Phi(u_n) = \tau/n$ (for a given $\tau > 0$) where Φ is the standard normal d.f.. Now $1 - \Phi(x) \sim \phi(x)/x$ as $x \to \infty$ where ϕ is the standard normal density so that $n\phi(u_n)/u_n \to \tau$. It follows by taking logs and rearranging that

(2.5) $\quad u_n^2 = 2 \log n - \log 2\pi - 2 \log u_n - 2 \log \tau$

and dividing by $\log u_n$ shows that $\log u_n = o(\log n)$. It then follows from (2.5) that $u_n^2 \sim 2 \log n$ or $u_n^2/(2 \log n) \to 1$, whence by taking logs we have

(2.6) $\quad \log u_n = \frac{1}{2} \log 2 + \frac{1}{2} \log \log n + o(1)$.

By substituting (2.6) in (2.5) and taking the square root we may obtain

$$u_n = (2 \log n)^{\frac{1}{2}} \left[1 - \frac{1}{2} \frac{\log \log n + \log 4\pi}{2 \log n} + \frac{\log \tau}{2 \log n} + o\left(\frac{\log \log n}{\log n}\right)^2 \right]$$

$$= b_n - \frac{\log \tau}{a_n} + o\left(\frac{1}{a_n}\right)$$

where

$$(2.7) \begin{cases} a_n = (2 \log n)^{\frac{1}{2}} \\ \\ b_n = (2 \log n)^{\frac{1}{2}} - \tfrac{1}{2}(2 \log n)^{-\frac{1}{2}}(\log \log n + \log 4\pi). \end{cases}$$

Hence, writing $\tau = e^{-x}$ in Theorem 2.5, we have

$$P\{a_n(M_n - b_n) + o(1)] \le x\} \to \exp(-e^{-x})$$

so that also $P\{a_n(M_n - b_n) \le x\} \to \exp(-e^{-x})$.

Thus in the normal case the asymptotic distribution for M is of Type I with the normalizing constants a_n, b_n given by (2.7).

(iv) Limiting Poisson distribution of Exceedances of u_n.

Let $\xi_1, \xi_2 \ldots$ be i.i.d. and suppose $\{u_n\}$ may be chosen to satisfy (2.3). Any j for which $\xi_j > u_n$ will be called an exceedance of u_n by $\{\xi_k\}$. The number N_n of exceedances of u_n by $\{\xi_k\}$ for $1 \le k \le n$ is binomial with parameters n, $1 - F(u_n)$. Since $n(1 - F(u_n)) \to \tau$ it follows that N_n is asymptotically Poisson, i.e. the following result holds.

THEOREM 2.6 If N_n is the number of exceedances of u_n(satisfying (2.3)) by the i.i.d. random variables $\xi_1, \xi_2 \ldots \xi_n$, then

$$(2.8) \qquad P\{N_n = r\} \to e^{-\tau}\tau^r/r!$$

for any fixed $r = 1,2,3\ldots$.

Thus the exceedances have this "asymptotic

Poisson character" whatever the form of F (provided (2.3) can be satisfied). This notion will be developed much more in Section 5 where the exceedances will be regarded as a point process.

(v) Asymptotic distribution of $M_n^{(r)}$, the rth largest of $\xi_1 \ldots \xi_n$.

Let $\{u_n\}$ be a sequence such that (2.3) holds, and hence by Theorem 2.6, (2.8) holds. But the event $\{M_n^{(r)} \leq u_n\}$ is pre<u>c</u>isely the event $\{N_n < r\}$ so that $P\{M_n^{(r)} \leq u_n\} = \sum_{s=0}^{r-1} P(N_n = s) \to e^{-\tau} \sum_{s=0}^{r-1} \tau^s/s!$. Summarizing this formally we have the following result.

<u>THEOREM 2.7</u> If $\{u_n\}$ satisfies (2.3) and $M_n^{(r)}$ is the rth largest of the i.i.d. random variables $\xi_1, \xi_2 \ldots \xi_n$ then

(2.9) $\qquad P\{M_n^{(r)} \leq u_n\} \to e^{-\tau} \sum_{s=0}^{r-1} \tau^s/s!$ as $n \to \infty$,

for all $r = 1, 2, 3\ldots$.

<u>COROLLARY:</u> If $\{u_n\}$ is a sequence such that $P\{M_n \leq u_n\} \to e^{-\tau}$ then (2.9) holds. (For then by Theorem 2.5, (2.3) holds.)

We may develop this line a little further – to obtain the asymptotic distribution of $M_n^{(r)}$ from that of M_n. Specifically if $P\{a_n(M_n - b_n) \leq x\} \to G(x)$ where G is non-degenerate (and hence of Type I, II, or III) then, if $0 < G(x) < 1$, by applying the above corollary with

$u_n = x/a_n + b_n$, $\tau = -\log G(x)$ we see that

(2.10) $\quad P\{a_n(M_n^{(r)} - b_n) \leq x\} \to G(x) \sum_{s=0}^{r-1} (-\log G(x))^s/s!$.

It is also easily seen (using continuity of the extreme value distribution G) that the limit is zero if $G(x) = 0$ and 1 if $G(x) = 1$. Thus the following result holds.

THEOREM 2.8 Suppose that $P\{a_n(M_n - b_n) \leq x\} \to G(x)$ is $n \to \infty$ where G is a non-degenerate d.f.. Then the asymptotic distribution of $M_n^{(r)}$ the rth largest of $\xi_1, \xi_2, \ldots \xi_n$ is given by (2.10). (If $G(x) = 0$ or 1 the right-hand side of (2.10) is to be taken as 0 or 1 respectively.)

3. DEPENDENCE RESTRICTIONS.

We turn now to consider dependent stochastic sequences $\xi_1, \xi_2 \ldots$. Although it is not always necessary to do so (cf. [8]), we shall concern ourselves with strictly stationary sequences, i.e. such that the finite dimensional distributions $F_{i_1 \ldots i_n}(x_1 \ldots x_n)$ of $\xi_{i_1} \ldots \xi_{i_n}$ have the property that $F_{i_1+p, \ldots i_n+p}(x_1 \ldots x_n) = F_{i_1 \ldots i_n}(x_1 \ldots x_n)$ for any choice of n, $i_1 \ldots i_n$, $x_1 \ldots x_n$.

Certain of the properties of Section 2 were generalized by G. S. Watson ([25]) to apply to "m-dependent" stationary sequences, and by R. M. Loynes ([19]) to "strongly mixing" sequences. The strong-mixing assumption is the usual one, i.e. requiring that

(3.1) $\quad |P(A \cap B) - P(A)P(B)| < g(k)$

for any events $A \in \sigma(\xi_1 \ldots \xi_m)$ $B \in \sigma(\xi_{m+k+1}, \xi_{m+k+2} \ldots)$ ($\sigma(\xi_1 \ldots \xi_m)$ denoting the σ-field generated by $\xi_1 \ldots \xi_m$ etc.) for any m, where $g(k)$ (the "mixing function") tends to zero as $k \to \infty$. (m-dependence requires $g(k) = 0$ for $k \geq m$.)

In particular Loynes obtained Gnedenko's Theorem under strong mixing, and showed that (2.4) holds under natural further conditions. In addition he showed that under such conditions - the asymptotic distribution of $M_n = \max(\xi_1 \ldots \xi_n)$ is the same as it would be if the ξ_i were i.i.d. with the same marginal d.f. as when dependent.

When the sequence is strongly mixing but does not necessarily satisfy further conditions, it is possible to obtain an asymptotic distribution of M_n which is different from that of the i.i.d. sequence with the same marginal d.f. Examples of this - and other related matters complementing the work of Loynes, are given in papers by O'Brien ([20], [21]).

For stationary normal sequences, conditions on the covariance function $r_n = E\xi_i\xi_{i+n}$ (taking $E\xi_i = 0$, $E\xi_i^2 = 1$) are more natural than mixing conditions. It has been shown by S. M. Berman ([1]) that the same asymptotic limit holds for M_n in this case as for an i.i.d. normal sequence (i.e. a Type I limit with normalizing constants given by (2.7)) if one of the following conditions holds.

(3.2) $\quad r_n \log n \to 0$ as $n \to \infty$, or $\sum_{n=1}^{\infty} r_n^2 < \infty$.

(Mittal and Ylvisaker ([27]) have recently shown that if the conditions (3.2) do not hold a variety of limiting distributions are possible.)

It is possible to weaken the strong mixing condition and still obtain Gnedenko's Theorem and the other results indicated above. It will further turn out that this weakening provides results which apply to normal sequences under the conditions (3.2). To see how the strong mixing condition should be weakened, note first that not all events A, B in (3.1) are of interest. In fact we are clearly primarily interested in events of the form $\{\xi_1 \leq u, \xi_2 \leq u, \ldots \xi_n \leq u\}$ which suggests requiring a condition of the type (2.1) to hold for all $A = \{\xi_{i_1} \leq u \ldots \xi_{i_p} \leq u\}, B = \{\xi_{j_1} \leq u \ldots \xi_{j_q} \leq u\}$ where $i_1 < i_2 \ldots < i_p < j_1 < j_2 \ldots < j_q$ and $j_1 - i_p \geq k$. Put in terms of the finite dimensional distribution functions (and writing $F_{i_1 \ldots i_n}(u)$ for $F_{i_1 \ldots i_n}(u, u \ldots u)$) this would require

(D):

$$|F_{i_1 \ldots i_p j_1 \ldots j_q}(u) - F_{i_1 \ldots i_p}(u) F_{j_1 \ldots j_q}(u)| < g(k)$$

whenever $i_1 < i_2 \ldots < i_p < j_1 \ldots < j_q$, $j_1 - i_p \geq k$, $g(k) \to 0$ as $k \to \infty$. However even (D) is not sufficiently weak to deal with normal sequences satisfying the weak covariance conditions (3.2). We may obtain a suitable condition by modifying (D) so that it is essentially required to hold for a single sequence of interest, rather than for every fixed value of u. More precisely if (u_n) denotes a given fixed sequence we shall say that $D(u_n)$ holds if

(3.3) $(D(u_n))$:

$$|F_{i_1\ldots i_p j_1\ldots j_q}(u_n) - F_{i_1\ldots i_p}(u_n) F_{j_1\ldots j_q}(u_n)| < \alpha_{n,k}$$

whenever $i_1 < i_2 \ldots < i_p < j_1 \ldots < j_q$, $j_1 - i_p \geq k$, where $\alpha_{n,k}$ is non-increasing in k and where

(3.4) $$\lim_{n\to\infty} \alpha_{n,k_n} = 0$$

for some sequence $k_n \to \infty$ with $k_n/n \to 0$.

It is apparent that (D) implies $(D(u_n))$ for any sequence u_n (any sequence $k_n \to \infty$, $k_n/n \to 0$ may be chosen). However it will turn out that $D(u_n)$ is implied by either condition (3.2) in the normal case, for suitably chosen u_n.

We shall see that Gnedenko's Theorem holds under conditions of the type $D(u_n)$. However (as for strong mixing) an additional condition is required to obtain results such as (2.4) and the limit law for normal sequences. The following condition $D'(u_n)$ is a variant of one used by Watson [25] and Loynes [19] and will be appropriate here. Again let $[u_n]$ be a given real sequence. Then $D'(u_n)$ holds if

(3.5) $(D'(u_n))$:

$$\limsup_{n\to\infty} n \sum_{j=2}^{n} P\{\xi_1 > u_{nk}, \xi_j > u_{nk}\} = o(1/k) \text{ as } k \to \infty.$$

This condition involves the bivariate distribution of the $\{\xi_j\}$ and is easily seen to hold for example if the ξ_j are i.i.d. and u_n satisfies (2.3). It also holds (Section 6) if the ξ_j form a stationary normal sequence satisfying either condition (3.2), again if $\{u_n\}$ satisfies (2.3).

4. EXTREME VALUE THEORY FOR STATIONARY SEQUENCES

(i) Gnedenko's Theorem under $D(u_n)$.

The following generalization of Gnedenko's Theorem holds.

THEOREM 4.1 If $\{\xi_n\}$ is a stationary sequence, $M_n = \max(\xi_1 \ldots \xi_n)$ having a non-degenerate asymptotic distribution,

$$P\{a_n(M_n - b_n) \leq x\} \to G(x) \quad (a_n > 0)$$

and if $D(u_n)$ holds for $u_n = x/a_n + b_n$ (for each real x), then $G(x)$ is a Type I, II, or III extreme value distribution function.

The proof may be found in detail in [15]. Its main pattern is that used by Loynes [19] for the mixing case with, however, some essential differences. It will be useful to indicate the general lines of proof here, without detail.

As for the i.i.d. case, it is only necessary to show that if (2.1) holds for $k = 1$, then it holds for all $k = 1,2,3\ldots$ The following basic lemma uses $D(u_n)$ to "approximate by independence". In this $M(E)$ will denote $\max(\xi_j: j \in E)$ for any set E of integers, and a set $(i, i + 1 \ldots j)$ will be called an interval.

LEMMA 4.2 Let N,r,k be fixed and $D(u_n)$ hold for a given sequence (u_n). Let E_j be intervals, $j = 1\ldots r$, separated by at least k from each other. Then

$$|P\{\bigcap_{j=1}^{r} M(E_j) \le u_N\} - \prod_{j=1}^{r} P\{M(E_j) \le u_N\}| \le (r-1)\alpha_{N,k}$$

Lemma 4.2 is proved by a simple induction and is used in obtaining the following result:

LEMMA 4.3 Let n,m,k be positive integers. If $D(u_n)$ holds then writing $N = nk$, $I_1 = (1,2\ldots(n-m))$, $I_1^* = (n - m + 1,\ldots,n)$,

$$|P\{M_N \le u_N\} - P^k\{M_n \le u_N\}| \le (k + K)P\{M(I_1) \le u_N \le M(I_1^*)\}$$

for some constant K.

This is proved (along the lines used by Loynes) by dividing the first $N = nk$ integers into intervals $I_1, I_1^*, I_2, I_2^*, \ldots I_k, I_k^*$ of "lengths" $n-m$, m, $n-m$, m .... . $P\{M_N \le u_N\}$ is then approximated by $P\{\bigcap_{j=1}^{k} M(I_j) \le u_N\}$ which in turn is (by Lemma 4.2) approximated by $\prod_{j=1}^{k} P\{M(I_j) \le u_N\}$. Each term is finally approximated by $P\{M_n \le u_N\}$.

From this lemma it can be shown that $P\{M_N \le u_N\} - P^k(M_n \le u_N) \to 0$ for each k. If $D(u_n)$ holds with $u_n = \frac{x}{a_n} + b_n$ and $P\{a_n(M_n - b_n) \le x\} \to G(x)$ (i.e. if (2.1) holds for $k = 1$) then $P\{M_N \le u_N\} \to G(x)$ and hence $P^k\{M_n \le u_{nk}\} \to G(x)$ or

$P\{a_{nk}(M_n - b_{nk}) \le x\} \to G^{1/k}(x)$ for all $k = 1, 2...$ (i.e. 2.1 holds for all $k = 1, 2...$) as required to finish the proof of Theorem 4.1.

(ii) Convergence of $P\{M_n \le u_n\}$ under (2.3), $D(u_n)$, $D'(u_n)$.

It is shown in [19] that under strong mixing if $P(M_n \le u_n(\tau)) \to \psi(\tau)$ when $u_n(\tau)$ satisfies (2.3) for each τ then $\psi(\tau) = e^{-\alpha\tau}$ for some α with $0 \le \alpha \le 1$. It may similarly be shown that this applies if, instead of strong mixing, the conditions $D(u_n(\tau))$ hold for each τ. (This is seen simply by using the fact that $[1 - F(u_{2n}(\tau))] \sim (\tau/2)/n$ so that $P\{M_n \le u_{2n}(\tau)\} \to \psi(\tau/2)$ and thus, by Lemma 2.5 of [15], $\psi(\tau) = \psi^2(\tau/2)$ from which the exponential limit may be deduced as in [19]).

O'Brien ([20]) gives examples in which $\alpha < 1$. We are here particularly interested in the "usual" case $\alpha = 1$ which may be guaranteed by the condition $D'(u_n)$ of the last section. Specifically we obtain the following result, again generalizing one given in [19] under strong mixing. It should be noted that the limit is shown to exist under the stronger conditions (rather than assumed to exist and then shown to have a given form as above, when just $D(u_n)$ holds).

THEOREM 4.4 Let $\{\xi_n\}$ be a stationary sequence and let $D(u_n)$, $D'(u_n)$ hold, where $\{u_n\}$ satisfies (2.3) for some fixed $\tau > 0$. (i.e. $1 - F(u_n) = P\{\xi_1 > u_n\} \sim \tau/n$.) Then

$$P\{M_n \le u_n\} \to e^{-\tau} \text{ as } n \to \infty.$$

The proof of this result is given in [15] and will not be repeated here. It is based on the standard inequalities

$$\sum_i P\{\xi_i > u\} - \sum_{i<j} P\{\xi_i > u, \xi_j > u\} \le P\{\bigcup_i (\xi_i > u)\}$$

$$= P(M_n > u)$$

$$\le \sum_i P\{\xi_i > u\}.$$

(iii) Relations with the i.i.d. case.

We write $\hat{M}_n$ to denote the maximum of n i.i.d. r.v.'s with the same distribution function F as the terms of the stationary sequence $\{\xi_i\}$.

THEOREM 4.5 Suppose that $P\{\hat{M}_n \le u_n\} \to \theta$ for some sequence u_n and some θ, $(0 < \theta < 1)$, and that $D(u_n)$, $D'(u_n)$ hold for the stationary sequence $\{\xi_j\}$. Then $P\{M_n \le u_n\} \to \theta$.

PROOF: By Theorem (2.5) $1 - F(u_n) \sim \tau/n$ with $\tau = -\log \theta$, i.e. (2.3) holds with this $\{u_n\}$. Hence by Theorem 4.4

$$P\{M_n \le u_n\} \to e^{-\tau} = \theta.$$

Thus the same limit holds for the maximum in the dependent sequence as would apply if the random variables were i.i.d. with the same marginal distribution function. As a corollary we may consider the asymptotic distribution of M_n under the standard normalization.

THEOREM 4.6 If $P\{a_n(\hat{M}_n - b_n) \leq x\} \to G(x)$ (non-degenerate) then $P\{a_n(M_n - b_n) \leq x\} \to G(x)$ if $D(u_n)$, $D'(u_n)$ hold for $u_n = x/a_n + b_n$.

This follows at once from Theorem 4.5 for $0 < G(x) < 1$ and by continuity of G, at points where $G(x) = 0$ or 1.

(iv) Stationary normal sequences.

If $\{\xi_j\}$ is a (zero mean, unit variance) stationary normal sequence whose covariances $\{r_n\}$ satisfy either condition (3.2), then it is known (Berman, [1]) that

$$P\{a_n(M_n - b_n) \leq x\} \to \exp(-e^{-x})$$

where a_n and b_n are given by (2.7). This result also follows as a consequence of Theorem 4.6. For this limit applies to i.i.d. random variables (as shown in Section 2) and it follows by Lemma 4.3 of [15] that either condition (3.2) implies both $D(u_n)$, $D'(u_n)$ when $u_n = x/a_n + b_n$, in this particular case.

The ultimate amount of work in this route to the asymptotic distribution in the normal case is not less than that of the original proof in [1]. However it does indicate the satisfactory nature of the D, D' conditions, since they are implied by the very weak conditions (3.2) when the sequence is normal.

5. EXCEEDANCES OF HIGH LEVELS BY A STATIONARY SEQUENCE, AND THE DISTRIBUTION OF r^{TH} LARGEST VALUES.

(i) Exceedances of high levels.

As previously noted, any point j where ξ_j exceeds u will be called an "exceedance of the level u" by the sequence $\{\xi_j\}$. The exceedances form a point process i.e. a series of events occurring in "time" according to some probabilistic law. Suppose now that $\{u_n\}$ is a sequence of constants satisfying (2.3). As n increases, the probability of an exceedance becomes smaller and the exceedances consequently tend to become rarer. An obvious question is whether one may thus obtain a convenient "limiting point process". We have seen in the i.i.d. case that the number of exceedances in (1,2...n) has an asymptotic Poisson distribution and one may expect that this will be true for stationary sequences under appropriate assumptions. We shall indicate how this may be shown - and indeed that the sequence of point processes formed from the exceedances of u_n converges weakly - as random elements of a natural metric space - to a Poisson process.

First we make a time scale change to avoid degeneracy. Let η_n be a discrete-parameter process defined on the points $\frac{j}{n}$, $j = 0,1,2...$, by $\eta_n(j/n) = \xi_j$. Consider the point process consisting of exceedances of u_n by η_n and let $N_n(B)$ denote the number of such exceedances in the Borel set B. (i.e. the number of exceedances of u_n by ξ_j for $j \in nB$.) Thus while we "lose exceedances" by increasing u_n we gain them in a compensating way by this change of time scale.

N_n is a point process for each n - i.e. a random element in the space N of non-negative integer-valued Borel measures on the real line R. N becomes a metric space with the "vague topology" (cf. [13]) and we may therefore consider convergence in distributions of these random elements. (That is a sequence $\{\zeta_n\}$ of point processes converges in distribution to a point process ζ, $\zeta_n \xrightarrow{D} \zeta$ if $Ef(\zeta_n) \to Ef(\zeta)$ for every bounded, vaguely continuous real f on N. The following is a useful criterion (due to Kallenberg [13] and modified according to a remark of Kurtz [14]) for such convergence.

<u>THEOREM 5.1</u> (Kallenberg)

Let ζ_n, n = 1,2... be point processes on the positive real line and let ζ be a point process without multiple points and such that $\zeta(\{a\}) = 0$ a.s. for every fixed real $a \geq 0$. If
- (i) $P\{\zeta_n(B) = 0\} \xrightarrow{}_r P\{\zeta(B) = 0\}$ for all sets B of the form $\bigcup_1^r (a_i, b_i]$ ($a_1 < b_1 < a_2 ... < a_r < b_r$)
- (ii) $\lim_n \sup E\zeta_n(a,b] \leq E\zeta(a,b]$ for all finite a < b then $\zeta_n \xrightarrow{d} \zeta$.

In our case $\zeta_n = N_n$, and ζ is a Poisson process with parameter τ, so that e.g.

$$P\{\zeta(B) = r\} = e^{-\tau m(B)} [\tau m(B)]^r / r!$$

where B is any Borel set and m denotes Lebesgue measure.

We shall suppose that $D(u_n)$, $D'(u_n)$ both hold and thus by Theorem 4.4 $P\{M_n \leq u_n\} \to e^{-\tau}$. But $\{M_n \leq u_n\}$ is the same as $\{N_n(0,1]) = 0\}$ so that

$$P\{N_n((0,1]) = 0\} \to e^{-\tau} = P\{\zeta((0,1]) = 0\}$$

i.e. Condition (i) holds when $B = (0,1]$.

It may be shown from this that Condition (i) holds for intervals $(0,\alpha]$ and then intervals $(a,b]$ and finally for finite disjoint unions $\cup(a_i, b_i]$. (The details of this development may be found in [17].)

Thus Condition (i) holds. Condition (ii) is easy to verify since (writing $[x]$ for the integer part of x)

$$EN_n(a,b] = ([nb] - [na])(1 - F(u_n))$$
$$\sim n(b - a)\tau/n = \tau(b - a)$$
$$= E\zeta((a,b]).$$

Thus Theorem 5.1 applies and we obtain the following result.

THEOREM 5.2 Let $\{\xi_j\}$ be a stationary sequence such that $D(u_n)$, $D'(u_n)$ hold for $\{u_n\}$ satisfying (3.2). Let the point process N_n consist of those points j/n for which $\xi_j > u_n$ ($N_n(B)$ being the number of such points in the Borel set B.) Then $N_n \xrightarrow{D} \zeta$ where ζ is a Poisson process with parameter τ.

COROLLARY. Under the conditions of the theorem, if B is any Borel set whose boundary has Lebesgue measure zero ($m(\partial B) = 0$) then for any $r = 0,1,2\ldots$,

$$P\{N_n(B) = r\} \to e^{-\tau m(B)}[\tau m(B)]^r/r!$$

(with similar convergence for the joint distribution of any $N_n(B_1)...N_n(B_k)$).

This follows at once since the random variables $N_n(B)$ converge in distribution to $\zeta(B)$ when $N_n \xrightarrow{D} \zeta$ (cf. [13]), and similarly for joint distributions.

(ii) Asymptotic distribution of $M_n^{(r)}$

THEOREM 5.3 Let $\{\xi_j\}$ be a stationary sequence such that $D(u_n)$, $D'(u_n)$ hold for $\{u_n\}$ satisfying (2.3). Then

$$P\{M_n^{(r)} \leq u_n\} \to \sum_{s=0}^{r-1} e^{-\tau}\tau^s/s!$$

($M_n^{(r)}$ being the rth largest among $\xi_1 \ldots \xi_n$).

PROOF: This result follows simply from Theorem 5.2 by noting that $P\{M_n^{(r)} \leq u_n\} = P\{N_n((0,1]) < r\}$ (cf. Theorem 2.7).

Finally we may generalize Theorem 4.6 to obtain (2.10) for our stationary sequences. $\hat{M}_n$ will again denote the maximum of n i.i.d. random variables with the same distributions as ξ_1.

THEOREM 5.4 Let $\{\xi_i\}$ be a stationary sequence. Suppose that $P\{a_n(M_n - b_n) \leq x\} \to G(x)$ (non-degenerate) and that $D(u_n)$, $D'(u_n)$ hold for $u_n = x/a_n + b_n$. Then (2.10) holds, i.e.

$$P\{a_n(M_n^{(r)} - b_n) \leq x\} \to G(x) \sum_{s=0}^{r-1} (-\log G(x))^s/s!.$$

PROOF: $\{u_n\} = \{x/a_n + b_n\}$ satisfies (2.3) by Theorem 2.5. The result now follows by Theorem 5.3

with $\tau = -\log G(x)$ (where $0 < G(x) < 1$) and by continuity arguments where $G = 0$ or 1.

6. CONTINUOUS PARAMETER STATIONARY PROCESSES

In this section we consider a continuous parameter stationary process $\{\xi(t): t \geq 0\}$ (assumed to have continuous sample functions), and write $M(T) = \sup\{\xi(t): 0 \leq t \leq T\}$.

(i) Strongly mixing processes.

The definition of strong mixing given is easily adapted to the continuous parameter case requiring, for any t,

$$|P(A \cap B) - P(A)P(B)| < g(\tau)$$

for $A \in \sigma\{\xi_s: s \leq t\}$, $B \in \sigma\{\xi_s: s \geq t + \tau\}$ where $g(\tau) \to 0$ as $\tau \to \infty$. Gnedenko's Theorem may be extended as follows

THEOREM 6.1 Suppose the stationary process $\{\xi(t)\}$ is strongly mixing and $P\{a_T(M(T) - b_T) \leq x\} \to G(x)$, non-degenerate, for some families $a_T > 0$, b_T of constants. Then $G(x)$ has one of the three extreme value forms.

PROOF: Write $Z_i = \sup\{\xi(t): i - 1 \leq t \leq i\}$. Then $\{Z_i: i = 1,2...\}$ may be seen to be a stationary and strongly mixing sequence. Further, putting $T = n$, an integer, $M(n) = \max\{Z_i: 1 \leq i \leq n\}$, and

$$P\{a_n(M(n) - b_n) \leq x\} \to G(x)$$

so that by Gnedenko's Theorem for strongly mixing sequences $G(x)$ has one of the three extreme value forms.

Other partial results may be shown under strong mixing or D, D' type conditions. For example we may see under appropriate conditions, that

$$P\{M(n) \leq u_n\} \to e^{-\tau} \text{ if } P\{Z_1 > u_n\} \sim \tau/n$$

($Z_1 = \sup\{\xi(t): 0 \leq t \leq 1\}$) and whence
(6.1) $$P\{M(T) \leq u_T\} \to e^{-\tau}$$

if

(6.2) $$P\{Z_1 > u_T\} \sim \tau/T.$$

This may be made rigorous under sufficiently strong conditions. However natural, simple conditions need to be found to make the result satisfying and useful in the general context. For the normal case there are such conditions, as will be seen below.

(ii) Stationary normal processes.

Extreme values of stationary processes have been considered in detail by a number of authors, including S. M. Berman, J. Pickands, C. Qualls and H. Watanabe. Let $\{\xi(t)\}$ then be a process with (for convenience) zero mean, unit variance and covariance function $r(\tau) = E\xi(t)\xi(t + \tau)$. It is usually assumed that $r(\tau)$ has the following expansion as $\tau \to 0$.

(6.3) $$r(\tau) = 1 - C|\tau|^\alpha + o(|\tau|^\alpha)$$

for some α with $0 < \alpha \leq 2$, though the theory can be extended to apply to the case where a function $G(\tau)$

of slow "growth" is included as a multiplicative factor together with $|\tau|^\alpha$ (cf. [24]).

$\alpha = 2$ gives the "regular" case - where $\xi(t)$ has a quadratic mean derivative and where the mean number of "upcrossings" of any level per unit time, is finite. If $0 < \alpha < 2$, the sample functions are continuous, but the mean number of upcrossings of a level u in any interval is infinite, and there is no q.m. derivative. The case $\alpha = 1$ corresponds to the Ornstein-Uhlenbeck Process.

The development indicated under (i) above may be applied to stationary normal processes satisfying (6.3) and in so doing we may replace strong mixing by either of the conditions

$$(6.4) \qquad r(t) \log t \to 0 \text{ or } \int_0^\infty r^2(t) dt < \infty$$

(i.e. the continuous analogues of (3.2)). Indeed the following result holds.

<u>LEMMA 6.1</u> Suppose (6.3) and (6.4) are satisfied, for the stationary normal process $\xi(t)$ considered above. Then

$$P\{M(T) \le u_T\} \to e^{-\tau} \text{ as } T \to \infty$$

if u_T is chosen so that (6.2) holds, i.e.

$$P\{\sup(\xi(t): 0 \le t \le 1) > u_T\} \sim \tau/T.$$

This lemma may be obtained e.g. from the discussion in [22] or [3]. From these references we may also obtain the asymptotic form of

$P\{\sup(\xi(t): 0 \le t \le 1) > u\}$ as follows.

LEMMA 6.2 Under the conditions of Lemma 6.1, as $u \to \infty$,

$$P\{\sup(\xi(t): 0 \le t \le 1) > u\} \sim C^{1/\alpha} H_\alpha u^{2/\alpha} \phi(u)/u$$

where H_α is a certain constant and ϕ is the standard normal probability density function. Thus the conclusion $P(M(T) \le u_T) \to e^{-\tau}$ of Lemma 6.1 holds if u_T is chosen to satisfy

(6.5) $C^{1/\alpha} u_T^{2/\alpha} \phi(u_T)/u_T \sim \tau/T$ as $T \to \infty$.

By taking logarithms in (6.5) we may obtain, after some manipulation

(6.6) $u_T = (2 \log T)^{1/2} - (2 \log T)^{-1/2}\Big[\log \tau +$

$\left(\frac{1}{2} - \frac{1}{\alpha}\right) \log \log \tau + \log[(2\pi)^{1/2} C^{1/\alpha} 2^{(\alpha-2)/2\alpha} H_\alpha^{-1}]$

$+ o(\log T)^{-\frac{1}{2}}$

and Lemma 6.2 at once gives the following theorem.

THEOREM 6.3 Let the stationary normal process $[\xi(t)]$ satisfy (6.2) and (6.3). Then

$$P\{a_T(M(T) - b_T) \le x\} \to \exp(-e^{-x}) \text{ as } T \to \infty$$

where

$$a_T = (2 \log T)^{1/2}$$

$$b_T = (2 \log T)^{1/2} - (2 \log T)^{-1/2} \left[\left(\tfrac{1}{2} - \tfrac{1}{\alpha}\right)\right] \log \log T$$

$$+ \log \left[(2\pi)^{1/2} \ C^{1/\alpha} \ 2^{(\alpha-2)/2\alpha} \ H_\alpha^{-1}\right]$$

PROOF: Put $\tau = e^{-x}$ in (6.6) to give $u_T = x/a_T + b_T$ and apply Lemma 6.2.

The value of the constant H_α is known only for $\alpha = 1, 2$ ($H_1 = 1$ $H_2 = \pi^{-1/2}$). Its form is given in [22], and depends on α, but not otherwise on the process $\xi(t)$.

The derivation of Lemmas 6.1 and 6.2 naturally involves considerable calculation, and different approaches may be used (cf. [22], [3]). In [22] use is made of the "ϵ-upcrossings" of a level. $\xi(t)$ is said to have an upcrossing of u at t_o if $\xi(t) \leq u$ for $t_o - \delta < t \leq t_o$ and $\xi(t) \geq u$ for $t_o \leq t < t_o + \delta$, for some $\delta > 0$, and this is an ϵ-upcrossing if there are no other upcrossings in $(t_o - \epsilon, t_o)$. As noted before if $\alpha = 2$ in (6.3) the number of upcrossings in $0 \leq t \leq 1$ has a finite mean, whereas this is not so for $\alpha < 2$. However the number of ϵ-upcrossings (for any fixed $\epsilon > 0$) in any finite interval is a bounded random variable, which turns out to be useful in the case $\alpha < 2$.

The (ϵ-)upcrossings in the continuous case naturally replace the exceedances in the discrete case. It is also possible to show that their limiting distribution (for suitably chosen high levels) is Poisson (cf. [4], [2], [23]). In fact it has recently been shown by Lindgren et al. ([18]) that these (ε-)upcrossings converge weakly to a Poisson process. The proof is similar to that described here in the sequence case (Sec. 5) for exceedances. Further questions concerning the asymptotic distribution of rth largest maxima and closely related matters are also discussed in [18].

Finally we note that it has also been recently shown by Mittal and Ylvisaker ([27]), that if the conditions (6.4) do not hold, a variety of other limit laws are possible for the maximum.

REFERENCES

(Including some related works
not referred to in text)

[1] Berman, S. M. Limit theorems for the maximum term in stationary sequences. Ann. Math. Statist., 35 (1964) pp. 502-516.

[2] Berman, S. M. Asymptotic independence of the numbers of high and low level crossings of stationary Gaussian processes. Ann. Math. Statist., 42 (1971) pp. 927-945.

[3] Berman, S. M. Maxima and high level excursions of stationary Gaussian processes. Trans. Amer. Math. Soc., 160 (1971) pp. 67-85.

[4] Cramér, H. and Leadbetter, M. R. Stationary and Related Stochastic Processes, John Wiley and Sons, New York, (1967).

[5] De Haan, L. On regular variation and its application to the weak convergence of sample extremes. Math. Centre Tract, 32 (1970) Amsterdam.

[6] Deo, C. M. A note on strong mixing Gaussian sequences. Ann. Prob., 1 (1973) pp. 186-187.

[7] Feller, W. An Introduction to Probability Theory and its Applications 2. John Wiley and Sons, New York, (1966).

[8] Galambos, J. On the distribution of the maximum of random variables. Ann. Math. Statist., 43 (1972) pp. 516-521.

[9] Galambos, J. A general Poisson limit theorem of probability theory. Duke Math. J., 40 (1973) pp. 581-586.

[10] Gnedenko, B. V. Sur la distribution limite du terme maximum d'une série aléatoire, Ann. Math., 44 (1943) pp. 423-453.

[11] Gnedenko, B. V. and Kolmogorov, A. N. Limit Distributions for Sums of Independent Random Variables. Addison Wesley, New York, (1954).

[12] Gumbel, E. J. Statistics of Extremes. Columbia Univ. Press, New York, (1958).

[13] Kallenberg, O. Characterization and convergence of random measures and point processes. Z. Wahr. verw. Geb., 27 (1973) pp. 9-21.

[14] Kurtz, T. Point processes and completely monotone set functions. Tech. Rept. Dept. Math, Univ. of Wisconsin (1973).

[15] Leadbetter, M. R. On extreme values in stationary sequences. Z. Wahr. verw. Geb., 28 (1974) pp. 289-303.

[16] Leadbetter, M. R. Lectures on extreme value theory. Tech. Rept. Dept. Math. Stat., Lund Univ. Sweden, (1974).

[17] Leadbetter, M. R. Weak convergence of high level exceedances by a stationary sequence. Inst. of Stat. Mimeo Series No. 933, Univ. of N. C., (1974).

[18] Lindgren, G., de Maré, J., and Rootzén, H. Weak convergence of high level crossings and maxima for one or more Gaussian processes. Tech. Rept. 4 Dept. Math. Stat., Lund Univ., Sweden, (1974).

[19] Loynes, R. M. Extreme values in uniformly mixing stationary stochasitc processes. Ann. Math. Statist., 36 (1965) pp. 993-999.

[20] O'Brien, G. L. Limit theorems for the maximum term of a stationary process. Ann. Prob., (1974) pp. 540-543.

[21] O'Brien, G. L. The maximum term of uniformly mixing stationary processes. To appear in Z. Wahr. verw. Geb.

[22] Pickands, J. Asymptotic properties of the maximum in a stationary Gaussian process. Trans. Amer. Math. Soc., 145 (1969) pp. 75-86.

[23] Pickands, J. Upcrossing probabilities for stationary Gaussian processes. Trans. Amer. Math. Soc., 145 (1969) pp. 51-73.

[24] Qualls, C., and Watanabe, H. Asymptotic properties of Gaussian processes. Ann. Math. Statist., 43 (1972) pp. 580-596.

[25] Watson, G. S. Extreme values in samples from m-dependent stationary stochastic processes. Ann. Math. Statist., 25 (1954) pp. 798-800.

[26] Welsch, R. E. A weak convergence theorem for order statistics from strong-mixing processes. Ann. Math. Statist., 42 (1971) pp. 1737-1946.

[27] Mittal, Y., and Ylvisaker, D. Limit distributions for the maxima of stationary Gaussian processes. To appear in J. Stoch. Proc.

Some Dimension Results for Processes
with Independent Increments

by William E. Pruitt*
University of Minnesota

1. Introduction

The goal of this paper is to give an expository account of a few problems concerned with sample path properties of processes with stationary independent increments. Surveys which are far more comprehensive have been prepared recently by Fristedt [13] and Taylor [47]. The existence of these papers has been very beneficial and has greatly simplified the preparation of the present account. In this paper we will concentrate on some questions about Hausdorff dimensions of certain random sets determined by the sample functions of the process. Much of the recent work in this area has been done by John Hawkes and his ideas play a very important part in this exposition.

---

*This article is an extended version of a lecture presented at the Summer Research Institute on Statistical Inference for Stochastic Processes held at Indiana University July 31-August 9, 1974. The preparation of this paper was supported in part by the National Science Foundation.

Section 2 contains a brief discussion of processes with stationary independent increments and of Hausdorff measures. The dimension results for stable processes are in Section 3. Some indications of proofs are given with the emphasis being on the ideas rather than details. Many of the original proofs can be simplified by using recent results giving uniform bounds for $\dim X(B)$ where B is a small time set and this has been done. In fact, this unification and simplification of ideas was the main reason for writing this paper. Finally, the corresponding problems for general processes with stationary independent increments are discussed in Section 4 with the solutions being indicated where they are known and the open problems pointed out where the solutions have not yet been found.

2. Preliminaries

Let $X_t(=X_t(\omega))$, $t \in \mathbb{R}^1_+$, be a process with stationary independent increments, defined on a probability space $(\Omega, \mathcal{F}, P)$, and taking values in $\mathbb{R}^d$. We assume that $X_0 = 0$. Because of the stationary independent increments, the finite dimensional distributions of the process are determined by the distribution of X_t. This distribution must be infinitely divisible and its characteristic function is given by

$$E \exp\{i(u,X_t)\} = \exp\{-t\psi(u)\}, \quad u \in \mathbb{R}^d,$$

where ψ is called the **exponent** of the distribution and has the representation (see [15, pp. 272,3])

$$\psi(u) = i(a,u) + \frac{1}{2}Q(u) - \int [e^{i(u,y)} - 1 - \frac{i(u,y)}{1+|y|^2}] \, \nu(dy)$$

where $a \in \mathbb{R}^d$, $(,)$ is the usual inner product in $\mathbb{R}^d$, Q is a non-negative definite quadratic form, and ν is a Borel measure (the Lévy measure) on $\mathbb{R}^d \setminus \{0\}$ such that $\int \min(1, |y|^2) \nu(dy) < \infty$. The first term contributes to a deterministic drift (but a contribution to this may also come from the integral), the second gives a Gaussian component, and the integral gives the jumps of the process and is the part in which we will usually be interested. If ν satisfies the additional condition $\int \min(1, |y|) \nu(dy) < \infty$, then the last term in the integrand is integrable and can be incorporated with the drift term. We will assume that this has been done in this case and thus ψ will have the form

$$\psi(u) = i(a,u) + \frac{1}{2}Q(u) - \int [e^{i(u,y)} - 1] \nu(dy) \, .$$

The functions $X_t(\omega)$ considered as functions of t for fixed ω are called the **sample functions** of the process. The theory of Hunt processes (see [5, pp. 45,6]) guarantees that we can find a version of the process which has sample functions which are right continuous and have left limits everywhere with probability one and is also a strong Markov process. We will assume throughout that we are dealing with such a version.

A few special processes are of particular importance. If $\psi(u) = \frac{1}{2}|u|^2$, the corresponding process is called **Brownian motion**. The properties that we will study were usually done first for the case of Brownian motion. However, we will consider Brownian motion as a special case of the <u>stable</u> processes. These have exponents of the form

$$\psi(u) = i(a,u) + \lambda |u|^\alpha \int_{S_d} w_\alpha(u,\theta)\mu(d\theta),$$

where $\lambda > 0$, $0 < \alpha \leq 2$, μ is a probability measure on $S_d = \{\theta \in R^d : |\theta| = 1\}$, and

$$w_\alpha(u,\theta) = [1 - i\,\text{sgn}(u,\theta)\tan\tfrac{1}{2}\pi\alpha]\,|(\tfrac{u}{|u|},\theta)|^\alpha, \text{ if } \alpha \neq 1,$$

$$w_1(u,\theta) = |(\tfrac{u}{|u|},\theta)| + \tfrac{2i}{\pi}(\tfrac{u}{|u|},\theta)\log|(u,\theta)|.$$

The parameter α is called the index of the stable process. The case $\alpha = 2$ corresponds to Brownian motion (for $a = 0$ and an appropriate choice of λ) and for $\alpha = 1$ the processes are called Cauchy processes. If $a = 0$ for $\alpha \neq 1$, or if $\alpha = 1$ and μ has its center of mass at the origin, the process is called <u>strictly stable</u>. Our consideration of stable processes here will be restricted to the strictly stable processes. In this case,

$$\psi(u) = |u|^\alpha \eta(\tfrac{u}{|u|}),$$

where η is a function of the argument of u. It is an easy consequence of this that for any $r > 0$, $r^{-1/\alpha} X_{rt}$ is a version of the same process as X_t. This so called <u>scaling property</u> of stable processes is extremely useful. If $\psi(u) = \lambda |u|^\alpha$ for some $\lambda > 0$, the process is called a <u>symmetric stable process</u> of index α. We should point out here that in some respects Brownian motion is quite different from the other stable processes. Perhaps the most noticeable difference is that the sample functions of Brownian motion are continuous (with probability one) while the sample functions of the other stable processes are discontinuous.

Another class of processes with stationary independent increments which play an important role are the ones taking values in R^1 whose sample functions are non-decreasing. These processes are called **subordinators**. Since they are nonnegative, it is usually more convenient to use Laplace transforms instead of characteristic functions. They take the form

$$E \exp\{-uX_t\} = \exp\{-tg(u)\},$$

where $g(u)$ is called the **subordinator exponent** and is defined by

$$g(u) = au + \int_0^\infty (1 - e^{-uy})\nu(dy),$$

the Lévy measure ν now being concentrated on the positive half-line and satisfying $\int_0^1 y\nu(dy) < \infty$. If $g(u) = \lambda u^\alpha$ with $\lambda > 0$ and $0 < \alpha < 1$, the corresponding process is called a **stable subordinator** of index α.

Our reason for using Hausdorff measures is that they give us a means of comparing the sizes of different sets in R^d that have zero Lebesgue measure. We will now define these measures.

Let $h:[0,\infty) \to [0,\infty)$ be continuous, monotone increasing, and satisfy $h(0) = 0$. For $E \subset \mathcal{R}^d$, define

$$\Lambda_\delta^h(E) = \inf_{\mathcal{S}} \sum_i h(\text{diam } S_i),$$

where

$$\mathcal{S} = \{\{S_i\}: E \subset \bigcup_i S_i, \text{ diam } S_i < \delta \text{ for all } i\}.$$

Then $\Lambda^h(E) = \lim_{\delta \to 0} \Lambda_\delta^h(E)$ is an outer measure called the Hausdorff h-measure. We will be primarily interested in the class of functions $h^\alpha(s) = s^\alpha$ for $\alpha > 0$. We will abuse the notation slightly by using $\Lambda^\alpha(E)$ to denote the h^α-measure of E. If $\alpha = d$, then Λ^α is a constant multiple of Lebesgue measure on $\mathcal{R}^d$. (For this and other properties of Hausdorff measures, see [38]). By using this one parameter family of measures we can define a dimension for any set $E \subset \mathcal{R}^d$. The definition is

$$\dim E = \inf\{\alpha > 0: \Lambda^\alpha(E) = 0\} = \sup\{\alpha > 0: \Lambda^\alpha(E) = \infty\}$$

with the interpretation that the sup is to be zero if there are no values of α for which $\Lambda^\alpha(E) = \infty$. This is called the Hausdorff dimension of the set E and for any set E it is a real number in the interval $[0,d]$. If the set E has positive Lebesgue measure, then $\dim E = d$. Even if E has zero Lebesgue measure, it could have $\dim E = d$ if it is "big" enough. Any countable set has dimension zero.

It may be instructive to consider the Cantor set, C, in R^1. Recall that it is constructed by removing the interval $(\frac{1}{3},\frac{2}{3})$ from $[0,1]$; then removing the middle third of each of the two remaining intervals and so on. Thus, for any n, the Cantor set is contained in 2^n intervals each having length 3^{-n}. Now if $\alpha > \log 2/\log 3$ and n is large enough so that $3^{-n} < \delta$, then

$$\Lambda_\delta^\alpha(C) \le 2^n 3^{-n\alpha} = 3^{n(-\alpha+\log 2/\log 3)}$$

and so $\Lambda_\delta^\alpha(C) = 0$ and hence $\Lambda^\alpha(C) = 0$ and dim $C \le \log 2/\log 3$. With a little more work one can show that dim $C = \log 2/\log 3$ but we will omit this. The lower bound is harder because in order to show that $\Lambda_\delta^\alpha(C)$ is large, one needs to show that the sums are large for <u>all</u> covers in $\mathcal{S}$.

It can happen that $\Lambda^{\dim E}(E)$ is zero, positive and finite, or infinite. In case it is zero or infinity, sometimes it is possible to obtain finer distinctions between the sizes of sets by finding a different function h, i.e. not a power, such that $\Lambda^h(E)$ is positive and finite.

One property of the dimension that is quite useful (and which the reader may easily verify) is that if $E_n \nearrow E$, then dim $E_n \nearrow$ dim E. The corresponding property for decreasing sequences of sets is not true as may be easily seen by letting a sequence of intervals shrink down to a point. It is also easy to see that $\dim(A \cup B) = \max(\dim A, \dim B)$. Putting this together with the above result on increasing sequences gives $\dim(\cup A_i) = \sup_i \dim A_i$.

There is a basic property relating stable processes and Hausdorff dimension which has been of fundamental importance in the study of the sample functions of processes with independent increments. It states that if E is an analytic subset of R^d, then

(2.1) $\quad \alpha + \dim E < d \Rightarrow P\{X_t^{\alpha,d} \in E \text{ for some } t > 0\} = 0$,

(2.2) $\quad \alpha + \dim E > d \Rightarrow P\{X_t^{\alpha,d} \in E \text{ for some } t > 0\} > 0$,

where $X_t^{\alpha,d}$ is a symmetric stable process of index α in R^d (see [45, pp. 252,3]). Thus one can determine the dimension of a set in R^1 or R^2 if one knows which symmetric stable processes will not hit the set; conversely, if one knows the dimension of such a set, then one can tell which symmetric stable processes will fail to hit it. This doesn't work for $d \geq 3$ since no stable process will hit a set E with $\dim E < d - 2$ because the index of the stable process cannot exceed 2. However, it is often possible to determine the dimension of sets for $d \geq 3$ by similar techniques (see, e.g. [12]). A useful variant of (2.1) and (2.2) is that if E is an analytic subset of R_+^1, then the statements are still correct if $X_t^{\alpha,d}$ is replaced by T_t^{α} where T_t^{α} is a stable subordinator of index α [17, p. 93]. (Note that it is important in this case that E be a subset of the positive reals since $(-\infty,0)$ can never be hit by T_t^{α}). In fact, (2.1) and (2.2) are true for all (strictly) stable processes when $d = 1$ (provided E is a subset of the correct half-axis in the case of a subordinator), and (2.1) is known even for $d \geq 2$. However, the status of (2.2) is not known for general stable processes when $d \geq 2$.

3. Dimension Properties for Stable Processes

The Range. The first results we will consider are concerned with the range of the process. Let

$$R = R(\omega) = X(\mathfrak{R}^1_+) = \{x \in \mathfrak{R}^d : x = X_t \text{ for some } t \in \mathfrak{R}^1_+\} .$$

Then R is a random set in $\mathfrak{R}^d$ but it turns out that its Hausdorff dimension does not depend on ω if we restrict ω to an appropriate set having probability one. The theorem for stable processes is that

$$(3.1) \qquad \dim R = \min(\alpha, d) \qquad \text{a.s.}$$

This result is due to Taylor [41] for the case of Brownian motion, McKean [31] for the symmetric stable processes, and Blumenthal and Getoor [2] for general stable processes. Note that for Brownian motion in $\mathfrak{R}^1$ it is clear from the fact that the sample functions are continuous and unbounded that $R = \mathfrak{R}^1$ a.s. This is still true if $\alpha > d = 1$ but the result is not so trivial in this case. The result (3.1) for $\alpha \leq d$ will follow from Theorem 1 below.

Equation (3.1) has been generalized in two different ways. For $\alpha \leq d$, it turns out that $\Lambda^\alpha(R) = 0$ a.s. But for Brownian motion with $d \geq 3$ if we change from the function s^2 to the function $h(s) = s^2 \log|\log s|$, then $0 < \Lambda^h(R) < \infty$ a.s. This result is due to Ciesielski and Taylor [6]. Similar results are available for planar Brownian motion

([11], [36], and [44]) and for the stable processes ([37], [48], [46], and [35]). We will not pursue this type of generalization here, although there are open problems in this area for general processes with stationary independent increments.

The other generalization of (3.1) is obtained by considering a small time set B, i.e. $B \subset \mathbb{R}_+^1$, and relating the dimension of the image of B under X, i.e. X(B), to the dimension of B. The theorem here for stable processes is that for $B \subset \mathbb{R}_+^1$,

(3.2) dim X(B) = min(α dim B,d)   a.s.

This was proved first for Brownian motion by McKean [30] and then by Blumenthal and Getoor [1], [2] for the other stable processes. The next result we will discuss is very similar to (3.2) but in a certain sense very much stronger. In order to make the distinction clearer, let us first look at (3.2) more carefully. It states that for every $B \subset \mathbb{R}_+^1$ there is an exceptional set Ω_B, having probability measure zero, such that if $\omega \notin \Omega_B$ then the equality in (3.2) holds. The question we want to examine now is whether we can find a single exceptional set having measure zero which works for all B. This is not clear since the union of the Ω_B's is an uncountable union. The somewhat surprising result is that this can be done if $\alpha \le d$. This is the content of Theorem 1 below. This was proved by Kaufman [26] for planar Brownian motion and by Hawkes and Pruitt [19] for the other strictly stable processes with $\alpha \le d$. However, this single exceptional set cannot be found if $\alpha > d$ and we will discuss an example which shows this later in the section. Before beginning our discussion of Theorem 1, we should point out that the interest in this generalization of (3.2) is not merely academic, as it might seem, but rather with a view to applications. We will indicate some of these applications later in this section but the idea is that it is quite useful to have a result like (3.2) when the set B can also be random. This, of course, is not allowed in (3.2) but Theorem 1 allows for this possibility.

For the proof of Theorem 1 we will need two covering lemmas. The first is due to Hawkes and Pruitt [19, p. 280]. The second is modeled after Lemma 3 of Hawkes [16]. Since the proofs are similar we will only prove the second one.

Lemma 1 (Covering Principle I). Let $\{t_n\}$ be a sequence of positive real numbers with $\sum_n t_n^p < \infty$ for some $p > 0$, and let C_n be a class consisting of N_n intervals of length t_n where $\log N_n = O(1)|\log t_n|$. If $\{\theta_n\}$ is a sequence of positive real numbers such that for some $\delta > 0$ we have

(3.3) $\quad P\{\sup_{0 \le s \le t_n} |X_s| \ge \theta_n\} = O(1) t_n^\delta$

then there exists a positive integer k such that, with probability one, for sufficiently large n, $X(I)$ can be covered by k spheres of radius θ_n whenever I is in C_n.

Lemma 2 (Covering Principle II). Let $\{\theta_n\}$ be a sequence of positive real numbers with $\sum_n \theta_n^p < \infty$

for some $p > 0$, and let C_n be a class consisting of N_n sets in $\mathbb{R}_d$ of diameter θ_n where $\log N_n = O(1)|\log \theta_n|$. If $\{t_n\}$ is a sequence of positive real numbers such that for some $\delta > 0$ we have

(3.4) $\qquad P\{\inf_{t_n \leq s < \infty} |X_s| \leq \theta_n\} = O(1)\theta_n^\delta$

then there exists a positive integer k such that, with probability one, for sufficiently large n, $\{t: X_t \in C\}$ can be covered by k intervals of length t_n whenever C is in C_n.

Proof. Fix a set $C \in C_n$ and define a sequence $\{\tau_j\}$ of stopping times by letting τ_0 be the first hitting time of C and for $j \geq 1$,

$$\tau_j = \begin{cases} \inf\{t \geq \tau_{j-1} + t_n : X_t \in C\}, & \text{if } \tau_{j-1} < \infty, \\ \infty, & \text{if } \tau_{j-1} = \infty, \end{cases}$$

with the understanding that the infimum is ∞ if C is not hit after $\tau_{j-1} + t_n$. Note that if $\tau_k = \infty$, then the intervals $[\tau_j, \tau_j + t_n]$, $j = 0, 1, \ldots, k-1$, will cover $\{t: X_t \in C\}$. Thus $P\{\{t: X_t \in C\}$ cannot be covered by k intervals of length $t_n\}$

$$\leq P\{\tau_k < \infty\} = E\{P\{\tau_k < \infty | \mathcal{F}_{\tau_{k-1}}\}\}.$$

The conditional probability is zero on $\{\tau_{k-1} = \infty\}$; on $\{\tau_{k-1} < \infty\}$, it follows from the strong Markov property that this conditional probability is equal to

$$P^{X_{\tau_{k-1}}}\{X_t \in C \text{ for some } t \geq t_n\}.$$

Now note that since X_t is right continuous $X_{\tau_{k-1}} \in \overline{C}$ whenever $\tau_{k-1} < \infty$. Also for any $x \in \overline{C}$,

$$P^x\{X_t \in C \text{ for some } t \geq t_n\} = P\{X_t \in C - x \text{ for some } t \geq t_n\}$$
$$\leq P\{|X_t| \leq \theta_n \text{ for some } t \geq t_n\}$$

since $C - x$ is contained in the sphere of radius θ_n about the origin. Thus

$$P\{\tau_k < \infty\} \leq P\{|X_t| \leq \theta_n \text{ for some } t \geq t_n\} \cdot P\{\tau_{k-1} < \infty\},$$

and so by induction,

$$P\{\tau_k < \infty\} \leq (P\{|X_t| \leq \theta_n \text{ for some } t \geq t_n\})^k = O(1)\theta_n^{k\delta},$$

the last bound being a consequence of (3.4). Thus the event that there exists some $C \in C_n$ such that $\{t: X_t \in C\}$ cannot be covered by k intervals of length t_n has probability $O(1)N_n\theta_n^{k\delta}$. It now follows from the bound on N_n that this is summable for k large enough and so an application of the Borel Cantelli lemma completes the proof.

Theorem 1. Let X_t be any strictly stable process of index α, $\alpha \leq d$, in $\mathcal{R}^d$. Then

$$P\{\dim X(B) = \alpha \dim B \text{ for all } B\} = 1.$$

Proof. First we prove the lower bound for $\alpha < d$. It is known [24] in this case that there is a constant c such that if S is any sphere in $\mathcal{R}^d$ of radius a,

(3.5) $\quad P\{X_t \in S \text{ for some } t \geq T\} \leq c\left(\dfrac{a}{T^{1/\alpha}}\right)^{d-\alpha}$.

(This is most easily proved if one uses a little potential theory. However, it can also be obtained fairly readily by elementary methods by making use of the scaling property). Fix a positive integer m and let C_n consist of all cubes in $\mathcal{R}^d$ with edge length 2^{-n} and vertices at points of the form $(j_1 2^{-n}, \ldots, j_d 2^{-n})$ with the j_k integers and $|j_k| \leq 2m2^n$. Now take $\gamma < \alpha$ and let $\theta_n = \sqrt{d} \, 2^{-n}$, $t_n = 2^{-n\gamma}$; it follows from (3.5) that (3.4) is satisfied. Then it is a consequence of the second covering principle that there is an integer k such that for n sufficiently large $\{t: X_t \in C\}$ can be covered by k intervals of length t_n for all $C \in C_n$. If B is any set in $\mathcal{R}_+^1$, let

$$B_m = \{t: X_t \in X(B) \text{ and } |X_t| \leq m\}.$$

Then $B_m \nearrow B' \supset B$, where $B' = \{t: X_t \in X(B)\}$. Now if $\rho > \dim X(B_m)$ and $\epsilon > 0$ we cover $X(B_m)$ with sets S_i of diameters D_i less than ϵ and so that $\Sigma D_i^\rho < \epsilon$. With n_i chosen so that $2^{-n_i - 1} < D_i \leq 2^{-n_i}$ we have each S_i contained in 2^d cubes of C_{n_i}. Thus, for ϵ sufficiently small, $\{t: X_t \in S_i\}$ can be covered by $k2^d$ intervals of length $2^{-n_i \gamma} < (2D_i)^\gamma$. This gives a cover of B_m with a small ρ/γ sum. Thus we have $\dim B_m \leq \rho/\gamma$; letting $\rho \to \dim X(B_m)$, we see that

$$\dim X(B) \geq \dim X(B_m) \geq \gamma \dim B_m ,$$

with probability one, simultaneously for all sets $B \subset \mathcal{R}_+^1$. Now we let $\gamma \to \alpha$ through a countable sequence and m tend to infinity to complete the proof of the lower bound for $\alpha < d$.

The proof for $\alpha = d$ must necessarily be different. Kaufman [26] gives an elegant direct proof for the case $\alpha = d = 2$. We will give a sketch of Hawke's [16] method which depends on the case already proved. The idea is to let T_t be a stable subordinator of index ρ which is independent of X_t and $Y = X \circ T$. It is straightforward to check that Y is a stable process of index $\alpha\rho$. Then for a given time set B, let $C = \{t: T_t \in B\}$. Now, we have

$$\dim X(B) \geq \dim X(T(C)) = \dim Y(C) \geq \alpha\rho \dim C$$

so long as we avoid a fixed set of measure zero. This chain of inequalities is interesting since, while both extremes are random, one depends only on X while the other depends only on the independent process T. Thus, the inequality remains true if we let dim C take on its largest value (excluding some sets of measure zero) and we shall see in (3.8) that this value is $(\rho + \dim B - 1)/\rho$. Letting $\rho \to 1$ through a countable sequence completes the proof in this exceptional case.

The proof of the upper bound is similar using the first covering principle so we will only give the outline. By the scaling property and the asymptotic behavior of the tail of the stable distributions, we have

$$(3.6) \quad P\{\sup_{0\le s\le T} |X_s| \ge a\} = P\{\sup_{0\le s\le 1} |X_s| \ge aT^{-1/\alpha}\}$$
$$\le c(aT^{-1/\alpha})^{-\alpha}.$$

Then we take $t_n = 2^{-n}$, $\theta_n = 2^{-n/\gamma}$ for some $\gamma > \alpha$ where C_n consists of the intervals $[(j-1)2^{-n}, j2^{-n}]$ for $j \le 2m2^n$. For $\rho > \dim B_m$, where $B_m = B \cap [0,m]$, find a cover of B_m with a small ρ sum; this leads to a cover of B_m with sets in the C_n and then by the first covering principle to a cover of $X(B_m)$ with a small $\gamma\rho$ sum. Thus, we have $\dim X(B_m) \le \gamma\rho$. The proof is completed by letting $\rho \to \dim B_m$, then $\gamma \to \alpha$ through a countable sequence, and finally letting $m \to \infty$. An interesting feature of this proof is that in proving each inequality one is able to start with a good cover of one set, either B or X(B), and from it construct a good cover of the other. Thus, the difficulty which one might expect, namely showing that all covers have a certain property, is avoided.

Zero Sets and Occupation Time Sets. The next results we will consider are concerned with the zero set of the process, i.e. $Z = \{t: X_t = 0\}$. The (strictly) stable processes will hit points only if $\alpha > d$ so that $Z = \{0\}$ a.s. if $\alpha \le d$. For $\alpha > d$ it was shown by Taylor [42] for the case of Brownian motion and by Blumenthal and Getoor [4] for the symmetric stable processes that $\dim Z = 1 - \alpha^{-1}$ a.s. We can use this result to give an example which shows that Theorem 1 cannot be valid for $\alpha > d$. Recall that the assertion of the theorem is that once a single exceptional set has been deleted then $\dim X(B) = \alpha \dim B$ for all B and all ω. Thus we could even let B depend on ω. But if we let B = Z, then X(B) = {0} so that dim X(B) = 0 while α dim B = $\alpha - 1 > 0$.

The zero set has a natural generalization to the occupation time set S of a given set $E \subset \mathcal{R}^d$, i.e.

$$S = \{t: X_t \in E\} .$$

The dimension result here, due to Hawkes [17], is that if $\alpha \geq d$, then

(3.7) $\dim S = (\alpha + \dim E - d)/\alpha$ a.s.

Note that if E = {0}, then S = Z and we have the earlier result about the zero set. If E is large enough we know it may be hit even if $\alpha < d$ and then S will be non-empty. It turns out to be the case that if dim E $\geq$ d - α, (3.7) is still true in a certain sense even when $\alpha < d$. The upper bound for dim S is valid but the lower bound may fail since X may not hit E and so S may even be empty. What one might hope for in this case is to have (3.7) hold conditionally given that $S \neq \emptyset$. This is known in some cases; see Theorems 2 and 5 of [17]. A useful version that can be stated in some generality for $\alpha < d \leq \alpha + \dim E$ is

(3.8) $\sup\{\theta: P\{\dim S > \theta\} > 0\} = (\alpha + \dim E - d)/\alpha$.

This is true in general if d = 1 although E must be a subset of the appropriate half-axis if X is a subordinator. For $d \geq 2$, it is only known for the symmetric stable processes. Note that

(3.9) $\dim S \le (\alpha + \dim E - d)/\alpha$ a.s.

is a consequence of (3.8). The inequality (3.9) is true in general, even for $d \ge 2$.

The idea of the proof of both (3.7) and (3.8) is to compose X with an independent stable subordinator T and observe that T will hit S iff $X \circ T$ will hit E. To avoid some technical details, consider $d = 1$ and X symmetric. Choose ρ so that $\alpha\rho + \dim E < 1$ and let T have index ρ; then $X \circ T$ will not hit E by (2.1) and hence T will not hit S. Thus, for almost all ω, $\dim S \le 1 - \rho$ by the subordinator version of (2.2). Letting $\rho \to (1 - \dim E)/\alpha$ through a countable sequence gives the upper bound. The lower bound is proved similarly starting with $\alpha\rho + \dim E > 1$. The reason that the lower bound in (3.8) is weaker is that while $X \circ T$ hitting E with positive probability does imply that T will hit S with positive probability, it may happen that for ω in a set of positive measure $S(\omega)$ will not be hit by T. This means that $\dim S$ may be smaller for some ω's but still leads to the result (3.8). The stronger result when $\alpha \ge d$ is obtained by showing that certain events having positive probability must actually have probability one in this case. The reason that (3.8) is not known in general for $d \ge 2$ is because of the uncertain status of (2.2) for general stable processes in $\mathcal{R}^d$, $d \ge 2$, that was mentioned earlier.

Intersections with Fixed Sets. A related result is concerned with the intersection of the range R of the process X with a fixed subset E of $\mathbb{R}^d$. If $\alpha > d$, we have already pointed out that $R = \mathbb{R}^1$ a.s. so that $R \cap E = E$ a.s. For $\alpha \leq d$ we can obtain information about the dimension of $R \cap E$ by applying Theorem 1 to (3.7) and (3.8). The point is that with S the occupation time set of E as before, we have $R \cap E = X(S)$. Thus we have that if dim $E \geq d - \alpha$,

(3.10) $\quad \sup\{\theta: P\{\dim R \cap E > \theta\} > 0\} = \alpha + \dim E - d$

for the symmetric stable processes and for general ones if $d = 1$ except that for subordinators we again need $E \subset \mathbb{R}^1_+$. We also have dim $R \cap E \leq \alpha + \dim E - d$ in general and dim $R \cap E = \dim E$ if $\alpha = d$. Note that in the application of Theorem 1 we are using it for $X(S)$ where S is random so that we actually need the full strength of the theorem.

Multiple Points. A property of the sample functions that has aroused considerable interest is the existence of multiple points. A point in $\mathbb{R}^d$ is called a double point if it is visited at least twice and it is called a point of multiplicity n if it is visited at n or more different times. If $\alpha > d$, then points are hit with probability one so that all points are points of multiplicity n for all n. The interesting case is for $\alpha \leq d$. This problem was studied first for Brownian motion in a series of papers by Dvoretzky, Erdös, and Kakutani [7], [8], [9], [10] and then solved by Takeuchi [40] and Taylor [45], [46] for the stable processes with the cases for $d \geq 3$ completed by Fristedt [12]. The result is that for a (strictly) stable process of index α in $\Re^d$, $\alpha \leq d$, points of multiplicity n exist with probability one iff $n < d/(d - \alpha)$. The dimension of the set of n - multiple points is then $d - n(d - \alpha)$ a.s. Thus, for example, a stable process of index 73/74 in $\Re^1$ or one of index 73/37 in $\Re^2$ will have points of multiplicity 73 but not of multiplicity 74 and the dimension of the set of points of multiplicity 73 will be 1/74 for the process in $\Re^1$ or 1/37 for the process in $\Re^2$. For a symmetric stable process or any strictly stable process in $\Re^1$ we can sketch a proof which is based on the ideas we have already considered. For simplicity we will consider double points but the general idea should be clear. First watch the process run until a fixed time T. This gives a (random) set of dimension α by Theorem 1; its dimension is not affected if we translate it by X_T. Call this translated set E. Now start the process over at T, i.e. watch the process $Y_t = X_{T+t} - X_T$. Any point in the intersection of E and the range of Y is a double point for X. Thus, if E were not random, it would follow from (3.10) that for any $\theta < 2\alpha - d$ the dimension of the double points would exceed θ with positive probability. The randomness of E does not matter since E is independent of Y and the phrase "with positive probability" can be changed to "with probability one" by proving a zero-one law. In the other direction one obtains in the same way that those points which are hit before time T and also after time T have dimension no larger than $2\alpha - d$. But the union of these sets as T runs through the positive rationals will include the double points and still have dimension no larger than $2\alpha - d$.

<u>Collisions</u>. The final problem that we will consider for stable processes is concerned with collisions of two independent stable processes X_t and Y_t in $\mathbb{R}^1$ with indices α_1 and α_2 respectively. There is no harm in assuming $\alpha_1 \geq \alpha_2$. In order for there to be a collision we need to have $\alpha_1 > 1$. There are then two sets of interest:

(3.11) $S = \{t: X_t = Y_t\}$
$C = \{x \in \mathbb{R}^1 : x = X_t = Y_t \text{ for some } t\}.$

The set S can be viewed as the zero set of the process $X_t - Y_t$ or as the occupation time set of the diagonal for the process (X_t, Y_t) in $\mathbb{R}^2$. Neither process is stable except when $\alpha_1 = \alpha_2$ so questions about S really take us out of the context of the present section. The occupation time problem hasn't been studied in this generality anyway, but the zero set problem has and looking ahead to (4.3) we find that

$$\dim S = 1 - 1/\alpha_1 \quad \text{a.s.}$$

Since $C = X(S) = Y(S)$, we can obtain

(3.12) $\dim C = \alpha_2(1 - 1/\alpha_1) \quad \text{a.s.}$

when $\alpha_2 \leq 1$ by applying Theorem 1. Note that this gives another example which shows that Theorem 1 cannot be valid for $\alpha > d$ because if we try to apply it to $C = X(S)$ we get the wrong answer. The result (3.12) is also correct when $\alpha_2 > 1$ even though the present proof doesn't work. The results on collisions are due to Jain and Pruitt [25] for $\alpha_2 > 1$ and Hawkes [16] for $\alpha_2 \leq 1$.

4. Dimension Properties for Processes with Stationary Independent Increments

In the results that we have described for the stable processes, the index α of the process has played a major role. Thus, when dealing with more general processes, we will need a parameter that will take the place of this index. This idea was first studied by Blumenthal and Getoor in a fundamental paper [3]. It turned out that different indices were needed for different problems and they defined four indices in their original paper. Two others have been defined since. For most of the indices there are several equivalent definitions. We will define them in terms of the exponent ψ of the distribution when this is possible. It will be assumed that there is no Gaussian part to the process, i.e. $Q = 0$, and that if $\int \min(1, |y|) \nu(dy) < \infty$, then the last term in the integral has been combined with the drift term and this combined drift term is zero. (If the drift term is left in it will dominate the random part of the process for t small). Note that, with two minor exceptions to be noted below, all of these indices are equal to the index α of the stable process when the process under consideration is stable.

Blumenthal and Getoor introduced the upper index

$$\beta = \inf\{\xi \geq 0: |u|^{-\xi} \text{Re } \psi(u) \to 0 \text{ as } |u| \to \infty\},$$

and two lower indices

$$\beta'' = \sup\{\xi \geq 0 : |u|^{-\xi} \operatorname{Re} \psi(u) \to \infty \text{ as } |u| \to \infty\},$$

$$\beta' = \sup\{\xi \geq 0 : \int |u|^{\xi-d} \frac{1 - \exp\{-\operatorname{Re} \psi(u)\}}{\operatorname{Re} \psi(u)} du < \infty\}.$$

They also introduced a lower index σ for subordinators defined in terms of the subordinator exponent g by

$$\sigma = \sup\{\xi \geq 0 : u^{-\xi} g(u) \to \infty \text{ as } u \to \infty\}.$$

The index γ introduced by Pruitt [34] has not been defined in general in terms of the exponent ψ. However, under the fairly weak condition that $\operatorname{Re} \psi(u) \geq 2 \log |u|$ for large $|u|$,

$$\gamma = \sup\{\xi < d : \int |u|^{\xi-d} \operatorname{Re} \frac{1 - \exp\{-\psi(u)\}}{\psi(u)} du < \infty\}.$$

In the case of a stable process, $\gamma = \min(\alpha, d)$. Finally the index b was introduced by Hawkes [18] for processes taking values in R^1 only. It is defined by

$$b^{-1} = \inf\{\xi : \xi \in [0,1] \text{ and } \int (1 + \operatorname{Re} \psi^\xi(u))^{-1} du < \infty\}$$

where $b = 1$ in case the formula asks for the infimum of the empty set. In the stable case, $b = \max(1, \alpha)$.

Generally the various indices can all be different except that in the case of a subordinator $\gamma = \sigma$. There are several inequalities known, however, and we will list these now.

$\beta'' \leq \beta' \leq \beta \leq 2$, $\min(\beta',d) \leq \gamma \leq \min(\beta,d)$

$\beta > 1$ implies $\max(1,\beta') \leq b \leq \beta$, $\beta \leq 1$ implies $b = 1$.

In the remainder of the paper we will discuss briefly the same problems that were discussed for stable processes in the last section indicating both the known results and the open problems.

The Range. The basic result here is that

$$\dim R = \gamma \quad \text{a.s.}$$

This was proved first for subordinators by Horowitz [23] and in the general case by Pruitt [34]. Some results are known concerning functions h which make $\Lambda^h(R)$ positive and finite in the case of subordinators [14] and of processes having stable components [35]. For the sets $X(B)$, Blumenthal and Getoor [3] obtained the upper bound

(4.1) $\quad \dim X(B) \leq \min(\beta \dim B, d) \quad$ a.s.

for $\beta < 1$. The restriction $\beta < 1$ was later removed by Millar [32]. Blumenthal and Getoor also obtained lower bounds, but here the situation is a little more complicated:

(4.2) $\quad \beta' \leq d$ implies $\dim X(B) \geq \beta' \dim B \quad$ a.s. ,

$\qquad \beta' > d$ implies $\dim X(B) \geq \min(\beta'' \dim B, d)$ a.s.

They also raised the question in [3] of whether the formula dim X(B) = dim X[0,1] · dim B was true in general. By considering a process X = (U,V) with two independent stable components in R^1, Hendricks [20] found the first example of a process for which this formula could fail. Hendricks [21] also found the dimension of X(B) for a general process with stable components. For the case with two one dimensional components with indices α_1, α_2 and $\alpha_1 \geq \alpha_2$, he showed that dim X(B) = α_1 dim B a.s. if $\alpha_1 \leq 1$, while if $\alpha_1 > 1$

$$\dim X(B) = \begin{cases} \alpha_1 \dim B & \text{if } \dim B \leq \alpha_1^{-1} \\ 1 - \alpha_2 \alpha_1^{-1} + \alpha_2 \dim B & \text{if } \dim B > \alpha_1^{-1}. \end{cases}$$

The upper bound (4.1) has been shown to be uniform in B by Hawkes and Pruitt [19]. The proof still proceeds by using the first covering principle. To take the place of (3.6) one proves that for $\xi > \beta$, $0 < \eta < 1 - \beta/\xi$, and c a positive constant, there is a constant A such that

$$P\{\sup_{0 \leq s \leq T} |X_s| \geq cT^{1/\xi}\} \leq AT^\eta$$

for all T. On the other hand, no uniform lower bounds are known for dim X(B) except for the case when X is a subordinator. The difficulty comes in trying to find an upper bound for the delayed hitting probabilities to take the place of (3.5). We can give a simple example to show that the lower bound in (4.2) cannot be uniform. Let X = (U,V) where U, V are independent symmetric stable processes in R^1 with indices α_1 and α_2 respectively and $0 < \alpha_2 \leq 1 < \alpha_1$. Then it is clear that $\beta = \alpha_1$, $\beta'' = \alpha_2$, and a little work shows that $\beta' = 1 + \alpha_2(1 - 1/\alpha_1)$. Since $d = 2$, $\beta' \leq d$ is satisfied. Let B be the zero set of U. Then $X(B) = \{0\} \times V(B)$ and so by Theorem 1,

$$\dim X(B) = \dim V(B) = \alpha_2 \dim B$$

and $\alpha_2 < \beta'$. Hawkes [18] has another example which shows that there is really no hope for a uniform lower bound unless one assumes $\beta \leq d$. But the question is still open as to whether there is a uniform lower bound, e.g. β'' dim B, when one does assume $\beta \leq d$. The situation for subordinators is more satisfactory. Hawkes [18] has shown that σ dim B is a uniform lower bound for dim $X(B)$ when X is a subordinator. This bound had been previously obtained by Blumenthal and Getoor [3] for fixed time sets. There is an example in [19] of a subordinator with $\sigma < \beta$ such that for every $\xi \in (\sigma, \beta)$ there is a set B of positive dimension with dim $X(B) = \xi$ dim B a.s. Thus the bounds cannot be improved. Another interesting feature of this example is that it is possible to find two different time sets B_1, B_2 having the same dimension but with dim $X(B_1)$ and dim $X(B_2)$ being different with probability one. This is the first example where the dimension of $X(B)$ has depended on features of B other than its dimension.

<u>Zero Sets</u>. The general result here is due to Hawkes [18] who proves that

(4.3) $\dim Z = 1 - 1/b$ a.s.

We have already pointed out the usefulness of this result in one case in the last section.

There seems to have been no work at all on the related problems of occupation time sets and the intersection of the range with a fixed set in $\mathcal{R}^d$.

Multiple Points. Hendricks [22] has shown that if $X = (U,V)$ where U,V are independent stable processes in $\mathcal{R}^1$ of indices α_1, α_2 respectively with $1 < \alpha_2 < \alpha_1$ then points of multiplicity n exist for X with probability one iff $n < (\alpha_1 + \alpha_2)/(\alpha_1 + \alpha_2 - \alpha_1\alpha_2)$. The dimension of the set of n-multiple points is then

$$\min(2 - n(\alpha_1 + \alpha_2 - \alpha_1\alpha_2)/\alpha_1,\ 1 + \{\alpha_1 - n(\alpha_1+\alpha_2-\alpha_1\alpha_2)\}/\alpha_2)$$

a.s. The proof of this result is considerably more involved than the one indicated in the stable case and gives some idea as to the difficulties that may arise for general processes. The only work on multiple points for more general processes is for the case of double points. Orey [33] showed that if X is symmetric and $\beta'' > d/2$ then the sample functions have double points a.s. On the other hand, if X has bounded continuous densities and for some $\alpha < d/2$, $\mathrm{Re}\{1/\psi(u)\} \geq |u|^{-\alpha}$ for all sufficiently large u, then there are no double points a.s. (For a symmetric process, the hypothesis is equivalent to requiring $\beta < d/2$). Hendricks (personal communication) has pointed out that Orey's results can be improved somewhat. But there is still a class of processes for which the question of existence of double points is not settled.

Collisions. The question about the dimension of the collision time set of two processes X_t and Y_t (the set S of (3.11)) is a question about the zero set of $X_t - Y_t$ and this was discussed above. However, it is now more difficult to obtain the dimension of the collision set C from the dimension of S. This has been done only in the case where X is a (strictly) stable process of index $\alpha > 1$ in $\mathcal{R}^1$ and Y is a subordinator with lower subordinator index σ. In this case,

$$\dim C = \sigma(1 - 1/\alpha) \quad \text{a.s.}$$

This is proved by showing that the dimension of S is $1 - 1/\alpha$ and then showing that S is almost surely in a restricted class of sets $\mathcal{S}$ for which $\dim Y(E) = \sigma \dim E$ for all E in $\mathcal{S}$ with probability one (see [19]).

References

(1). R. M. Blumenthal and R. K. Getoor. Some theorems on stable processes. Trans. Amer. Math. Soc. 95, 263-273, (1960).

(2). R. M. Blumenthal and R. K. Getoor. A dimension theorem for sample functions of stable processes. Ill. J. Math. 4, 370-375, (1960).

(3). R. M. Blumenthal and R. K. Getoor. Sample functions of stochastic processes with stationary independent increments. J. Math. Mech. 10, 493-516, (1961).

(4). R. M. Blumenthal and R. K. Getoor. The dimension of the set of zeros and the graph of a symmetric stable process. Ill. J. Math. 6, 308-316, (1962).

(5). R. M. Blumenthal and R. K. Getoor. <u>Markov Processes and Potential Theory</u>, Academic Press, New York (1968).

(6). Z. Ciesielski and S. J. Taylor. First passage times and sojourn times for Brownian motion in space and the exact Hausdorff measure of the sample path. Trans. Amer. Math. Soc. 103, 434-450, (1962).

(7). A. Dvoretzky, P. Erdös and S. Kakutani. Double points of paths of Brownian motion in n-space. Acta Sci. Math. 12, 75-81, (1950).

(8). A. Dvoretzky, P. Erdös and S. Kakutani. Multiple points of paths of Brownian motion in the plane. Bull. Res. Council Israel Sect. F 3, 364-371, (1954).

(9). A. Dvoretzky, P. Erdös and S. Kakutani. Points of multiplicity c of plane Brownian paths. Bull. Res. Council Israel Sect. F 7, 175-180, (1958).

(10). A. Dvoretzky, P. Erdös, S. Kakutani, and S. J. Taylor. Triple points of Brownian motion in 3-space. Proc. Cambridge Philos. Soc. 53, 856-862, (1957).

(11). P. Erdös and S. J. Taylor. On the Hausdorff measure of Brownian paths in the plane. Proc. Cambridge Philos. Soc. 57, 209-222, (1961).

(12). B. E. Fristedt. An extension of a theorem of S. J. Taylor concerning the multiple points of the symmetric stable process. Z. Wahrscheinlichkeitstheorie 9, 62-64, (1967).

(13). B. E. Fristedt. Sample Functions of Stochastic Processes with Stationary, Independent Increments. Advances in Probability, Vol. 3, Marcel Dekker, New York, (1974).

(14). B. E. Fristedt and W. E. Pruitt. Lower functions for increasing random walks and subordinators, Z. Wahrscheinlichkeitstheorie 18, 167-182, (1971).

(15). I. I. Gikhman and A. V. Skorokhod. Introduction to the Theory of Random Processes, Saunders, Philadelphia, (1965).

(16). J. Hawkes. Some dimension theorems for the sample functions of stable processes. Indiana Univ. Math. J. 20, 733-738, (1971).

(17). J. Hawkes. On the Hausdorff dimension of the intersection of the range of a stable process with a Borel set. Z. Wahrscheinlichkeitstheorie 19, 90-102, (1971).

(18). J. Hawkes. Local times and zero sets for processes with infinitely divisible distributions. J. London Math. Soc., (to appear).

(19). J. Hawkes and W. E. Pruitt. Uniform dimension results for processes with independent increments. Z. Wahrscheinlichkeitstheorie 28, 277-288, (1974).

(20). W. J. Hendricks. Hausdorff dimension in a process with stable components--an interesting counterexample. Ann. Math. Statis. 43, 690-

694, (1972).

(21). W. J. Hendricks. A dimension theorem for processes with stable components. Ann. Probability 1, 849-853, (1973).

(22). W. J. Hendricks. Multiple points for a process in $\mathbb{R}^2$ with stable components, Z. Wahrscheinlichkeitstheorie 28, 113-128, (1974).

(23). J. Horowitz. The Hausdorff dimension of the sample path of a subordinator. Israel J. Math. 6, 176-182, (1968).

(24). N. Jain and W. E. Pruitt. The correct measure function for the graph of a transient stable process. Z. Wahrscheinlichkeitstheorie 9, 131-138, (1968).

(25). N. Jain and W. E. Pruitt. Collisions of stable processes. Ill. J. Math. 13, 241-248, (1969).

(26). R. Kaufman. Une propriété métrique du mouvement brownien. C. R. Acad. Sci. Paris. 268, 727-728, (1969).

(27). R. Kaufman. Brownian motion and dimension of perfect sets. Can. J. Math. 22, 674-680, (1970).

(28). P. Lévy. <u>Processus stochastiques et mouvements browniens</u>. Gauthier Villiers, Paris, (1948).

(29). P. Lévy. La mésure de Hausdorff de la courbe du mouvement brownien. Giorn. Ist. Ital. Attuari 16, 1-37, (1953).

(30). H. P. McKean. Hausdorff-Besicovitch dimension of Brownian motion paths. Duke Math. J. 22, 229-234, (1955).

(31). H. P. McKean. Sample functions of stable processes. Ann. Math. 61, 564-579, (1955).
(32). P. W. Millar. Path behaviour of processes with stationary independent increments. Z. Wahrscheinlichkeitstheorie 17, 53-73, (1971).
(33). S. Orey. Polar sets for processes with stationary independent increments. pp. 117-126, Markov processes and potential theory. Wiley, New York, (1967).
(34). W. E. Pruitt. The Hausdorff dimension of the range of a process with stationary independent increments. J. Math. Mech. 19, 371-378, (1969).
(35). W. E. Pruitt and S. J. Taylor. Sample path properties of processes with stable components. Z. Wahrscheinlichkeitstheorie 12, 267-289, (1969).
(36). D. Ray. Sojourn times and the exact Hausdorff measure of the sample path for planar Brownian motion. Trans. Amer. Math. Soc. 106, 436-444, (1963).
(37). D. Ray. Some local properties of Markov processes, Proc. Fifth Berkeley Symposium, Vol. 2, Pt. 2, University of California Press, Berkeley, (1967).
(38). C. A. Rogers. Hausdorff measures. Cambridge University Press, Cambridge, (1970).
(39). C. A. Rogers and S. J. Taylor. Functions continuous and singular with respect to a Hausdorff measure. Mathematika 8, 1-31, (1961).
(40). J. Takeuchi. On the sample paths of the symmetric stable processes in space. J. Math.

Soc. Japan 16, 109-127, (1964).

(41). S. J. Taylor. The Hausdorff α-dimensional measure of Brownian paths in n-space. Proc. Cambridge Philos. Soc. 49, 31-39, (1953).

(42). S. J. Taylor. The α-dimensional measure of the graph and the set of zeros of a Brownian path. Proc. Cambridge Philos. Soc. 51, 265-274, (1955).

(43). S. J. Taylor. On the connection between Hausdorff measures and generalized capacity. Proc. Cambridge Philos. Soc. 57, 524-531, (1961).

(44). S. J. Taylor. The exact Hausdorff measure of the sample path for planar Brownian motion. Proc. Cambridge Philos. Soc. 60, 253-258, (1964).

(45). S. J. Taylor. Multiple points for the sample paths of the symmetric stable process. Z. Wahrscheinlichkeitstheorie 5, 247-264, (1966).

(46). S. J. Taylor. Sample path properties of a transient stable process. J. Math. Mech. 16, 1229-1246, (1967).

(47). S. J. Taylor. Sample path properties of processes with stationary independent increments. **Stochastic Analysis.** (D. G. Kendall and E. F. Harding, ed.), Wiley, London, (1973).

(48). S. J. Taylor and J. G. Wendel. The exact Hausdorff measure of the zero set of a stable process. Z. Wahrscheinlichkeitstheorie 6, 170-180, (1966).

CONDITIONED LIMIT THEOREMS FOR RANDON WALKS[*]

by Donald L. Iglehart
Stanford University

1. Introduction and Summary

In this paper we shall discuss some recent results on conditioned limit theorems for random walks and indicate some open problems in this area.

The set-up for these problems is the classical one. Let $\{\xi_k : k \geq 1\}$ be a sequence of independent, identically distributed random variables with mean μ and variance σ^2. Next form the random walk $\{S_n : n \geq 0\}$ by setting $S_0 = 0$ and $S_n = \xi_1 + \ldots + \xi_n$, $n \geq 1$. Define the random function X_n by

$$X_n(t) = S_{[nt]}/\sigma n^{\frac{1}{2}}, \quad 0 \leq t \leq 1,$$

where $[x]$ is the greatest integer in x. The first

---

[*]This work was supported by National Science Foundation Grant GP-31392X2 and Office of Naval Research Contract N00014-67-A-0112-0031. Presented at the Summer Research Institute of the Institute of Mathematical Statistics, Indiana University, July 31, 1974.

weak convergence result, DONSKER (1951), says that when $\mu = 0$ and $0 < \sigma^2 < \infty$ $X_n \Rightarrow W$, where W is Brownian motion and $\Rightarrow$ denotes weak convergence in the Skorohod topology on $D[0,1] \equiv D$; see BILLINGSLEY (1968), page 137, for further discussion of this result. Since Donsker this result has been generalized in many directions; for example PROHOROV (1956), Theorem 3.1, treats triangular arrays, SKOROHOD (1956), attraction to stable processes, BILLINGSLEY (1968), Chapter 4, dependent variables, and McLEISH (1974) martingales and near-martingales. Thus it seems apparent that generalizations of Donsker's theorem are nearing the end of the road. However, once the door is opened to conditioned limit theorems for X_n an enormous array of interesting problems arises.

Our concern here is with the weak convergence of X_n conditioned on an event Λ_n, say, whose probability is converging to zero as n becomes large. In the last six years there has begun to take shape a literature associated with this problem. For the reader who may be interested in pursuing this area we list the following references: DWASS and KARLIN (1963), LIGGETT (1968, 1970a, b), BELKIN (1970, 1972), PORT and STONE (1971), KAIGH (1974) and IGLEHART (1974a,b). In this paper we shall restrict our attention to conditioning events involving the hitting time, T, of the set $(-\infty, 0]$ by the random walk:

$$T = \inf\{n > 0: S_n \leq 0\},$$

where the infimum of the empty set is taken to be $+\infty$. In particular, we shall consider conditioning X_n on either $\{T > n\}$ or $\{T = n\}$.

Before listing the specific problems to be discussed we mention an application of these results which provided our initial motivation. If W_n is the waiting time of the nth customer in a general single server queue, then it is well-known that $W_n = \max\{S_n - S_r : 0 \leq r \leq n\}$, $n \geq 0$; see [10], for example. The ξ_i's in this application are differences of service and interarrival times. In this context T is the number of customers served in the first busy period. Observe that $W_n = S_n$ on $\{T > n\}$. Thus conditioning on $\{T > n\}$ will yield limit theorems for the waiting time process, given that the first busy period has not ended. Similarly conditioning on $\{T=n\}$ provides results for exactly the first busy period.

In Section 2 we discuss recurrent random walk, $\mu = 0$, conditioned on either $\{T > n\}$ or $\{T=n\}$. Associated results for renewal processes and random partial sums are dealt with in Section 3. The case of random walks with a negative drift, $\mu < 0$, is discussed in Section 4. Finally, in Section 5 a few of the many open problems in this area are mentioned.

2. Recurrent Random Walk

Suppose now that $\{\xi_k : k \geq 1\}$ are the coordinate functions on the product space $(\Omega, \mathcal{F}, P)$ and that $\mu = E\{\xi_1\} = 0$. The first step to take when considering conditioned limit theorems is to obtain

the asymptotic behavior of the probability of the
conditioning events. To that end set $r_n = P\{T>n\}$
and $f_n = P\{T=n\}$. Then the asymptotic results
we need are contained in the following known lemmas.

(2.1) LEMMA [SPITZER (1960), Theorem 3.5]. <u>If</u>
$E\{\xi_1\} = 0$ <u>and</u> $0 < \sigma^2 < \infty$, <u>then</u>

$$r_n \sim cn^{-\frac{1}{2}} \quad \text{as} \quad n \to \infty$$

<u>where</u>
$$c = \pi^{-\frac{1}{2}} \exp\{ \sum_{k=1}^{\infty} k^{-1}(\tfrac{1}{2} - P[S_k > 0])\} .$$

(2.2) LEMMA [BOROVKOV (1970), Corollary 9]. <u>If</u>
$E\{\xi_1\} = 0$, $E|\xi_1|^3 < \infty$, <u>and</u> ξ_1 <u>is nonlattice or
integer-valued with span 1, then</u>

$$f_n \sim (c/2) n^{-3/2} \quad \underline{\text{as}} \quad n \to \infty.$$

Thus in both cases of conditioning, on either
$\{T > n\}$ or $\{T=n\}$, we are examining a microscopic
slice of the original probability space $(\Omega, \mathcal{F}, P)$.
If $\Lambda_n = \{T>n\}$, then we let $(\Lambda_n, \Lambda_n \cap \mathcal{F}, P_n)$ be the
trace of $(\Omega, \mathcal{F}, P)$ on Λ_n, where $\Lambda_n \cap \mathcal{F} = \{\Lambda_n \cap \mathcal{F} : F \in \mathcal{F}\}$
and $P_n(A) = P\{A\}/P\{\Lambda_n\}$ for $A \in \Lambda_n \cap \mathcal{F}$. Similarly,
if $\Theta_n = \{T=n\}$, let $(\Theta_n, \Theta_n \cap \mathcal{F}, Q_n)$ be the trace of
$(\Omega, \mathcal{F}, P)$ on Θ_n. Let $D \equiv D[0,1]$ be the space of
real-valued, right-continuous functions on $[0,1]$
having left limits and $\mathcal{D}$ be the σ-field of Borel
sets generated by the open sets of the Skorohod
J_1-topology. Then we define the random functions
$(X_n | T > n)$ and $(X_n | T = n)$ by

and
$$(X_n|T > n)(t,\omega) = X_n(t,\omega), \quad (t,\omega) \in [0,1] \times \Lambda_n$$
$$(X_n|T = n)(t,\omega) = X_n(t,\omega), \quad (t,\omega) \in [0,1] \times \Theta_n.$$

Before stating the conditioned limit theorems we must define the stochastic processes which play the role of Brownian motion, W, in Donsker's theorem. The first process, W^+, we have dubbed <u>Brownian meander</u>. This process first arose to our knowledge in BELKIN (1972), Theorem 3.1, and came up again in [11]. Let $\tau_1 = \sup\{t \in [0,1]: W(t) = 0\}$. Set $\Delta_1 = 1 - \tau_1$. Then

$$W^+(t) = |W(\tau_1 + t\Delta_1)|/\Delta_1^{\frac{1}{2}}, \quad 0 \le t \le 1.$$

It is a continuous, non-homogeneous Markov process and has transition density given by

(2.3) $P\{W^+(t) \in dy\}$
$$= p(0,0,t,y)dy = t^{-3/2} y \exp(-y^2/2t) |N| \cdot$$
$$\cdot (y/(1-t)^{\frac{1}{2}}) dy$$

for $0 < t \le 1$ and $y > 0$; for $0 < s < t \le 1$ and $x, y > 0$

(2.4) $P\{W^+(t) \in dy | W^+(s) = x\} = p(s,x,t,y)dy =$
$$g(t-s,x,y) \frac{|N|(y/(1-t)^{\frac{1}{2}})}{|N|(x/(1-s)^{\frac{1}{2}})} dy,$$

where $g(t,x,y) = (2\pi t)^{-\frac{1}{2}}[\exp\{-(y-x)^2/2t\} - \exp\{-(y+x)^2/2t\}]$ and $|N|(x) = (2/\pi)^{\frac{1}{2}} \int_0^x \exp(-u^2/2)du$, $x \geq 0$. For the derivation of this transition density see BELKIN (1972), p. 61. Note that the $P\{W^+(1) \leq x\} = 1 - \exp(-x^2/2)$, $x \geq 0$, which is the Rayleigh distribution.

The second process, W_0^+, is <u>Brownian excursion</u>. Let $\tau_2 = \inf\{t \geq 1 : W(t) = 0\}$. Set $\Delta_2 = \tau_2 - \tau_1$ and

$$W_0^+(t) = |W(\tau_1 + t\Delta_2)|/\Delta_2^{\frac{1}{2}}, \quad 0 \leq t \leq 1.$$

Brownian excursion is also a continuous, non-homogeneous Markov process with transition density given by

(2.5) $P\{W_0^+(t) \in dy\} = q(0,0,t,y)\,dy = \dfrac{2y^2 \, e^{-y^2/2t(1-t)}}{\sqrt{2\pi t^3 (1-t)^3}} dy$

for $0 < t < 1$ and $y > 0$; for $0 < s < t < 1$ and $x,y > 0$

(2.6) $P\{W_0^+(t) \in dy | W_0^+(s) = x\}$

$= q(s,x,t,y)\,dy = g(t-s,x,y)\left(\dfrac{1-s}{1-t}\right)^{3/2} \cdot \dfrac{ye^{-y^2/2(1-t)}}{xe^{-x^2/2(1-s)}} dy;$

see ITÔ-McKEAN (1965), p. 76 for this result.

Clearly the distribution of $W_0^+(0)$ and $W_0^+(1)$ are degenerate at 0. We note in passing that for $0 \leq t \leq 1$ and $r > 0$.

$$E\{[W_0^+(t)]^r\} = 2^{(r+2)/2} \Gamma(\tfrac{r+3}{2}) \pi^{-\tfrac{1}{2}} [t(1-t)]^{r/2}.$$

Next consider the sequence of random functions $\{(X_n | T > n) : n \geq 1\}$. Our goal is to show under appropriate conditions (those of Lemma 2.2) that $(X_n | T > n) \Rightarrow W^+$. The route taken is the standard one: first show convergence of the finite-dimensional distributions (f.d.d.'s) and then show tightness. Here we shall only indicate the main ideas behind the proof; see [11], Theorem 3.4, for complete details. First look at $(X_n(1) | T > n)$ and recall that $S_n = W_n$ on $\{T > n\}$. Thus for $x \geq 0$

$$(2.7) \quad P_n\{X_n(1) \leq x\} = r_n^{-1} P\{W_n/\sigma n^{\tfrac{1}{2}} \leq x, T > n\}$$

$$= r_n^{-1} P\{W_n/\sigma n^{\tfrac{1}{2}} \leq x\} - r_n^{-1} \cdot \sum_{k=1}^{n} P\{W_n/\sigma n^{\tfrac{1}{2}}, T = k\}.$$

Since T is an almost everywhere finite, optional random variable and $W_T = 0$, we can write (2.7) after summation by parts as

$$(2.8) \quad P_n\{X_n(1) \leq x\} = P\{W_n/\sigma n^{\tfrac{1}{2}} \leq x\} - \sum_{k=1}^{n} (f_k/r_n)$$
$$\times [P\{W_{n-k}/\sigma n^{\tfrac{1}{2}} \leq x\} - P\{W_n/\sigma n^{\tfrac{1}{2}} \leq x\}].$$

From (2.1) and (2.2) we know the asymptotic behavior of r_n and f_n. If we let $M_n = \max\{S_r : 0 \leq r \leq n\}$, then W_n and M_n are known to have the same distribution; cf. FELLER (1971), page 198. Also when $\mu = 0$ and $0 < \sigma^2 < \infty$, $M_n/\sigma n^{\frac{1}{2}} \Rightarrow |N|$, the positive normal with density $(2/\pi)^{\frac{1}{2}} \exp\{-x^2/2\}$ for $x \geq 0$. So we know the limit of the first term on the right-hand side of (2.8). The sum on the right-hand side of (2.8) can be shown to be the Riemann approximating sum for a certain integral. To do this however, we required the difficult result of NAGAEV (1969), page 443: if $E|\xi_1|^3 < \infty$ then

$$\sup_{x \geq 0} |P\{M_n/n^{\frac{1}{2}} \leq n\} - |N|(x)| \leq Kn^{-\frac{1}{2}}$$

for all $n \geq 1$, where K is a finite, positive constant. After some asymptotic analysis these facts applied to (2.8) yield

(2.9) $$(X_n(1)|T > n) \Rightarrow W^+(1).$$

Before continuing with the weak convergence argument it is amusing to compare conditioning on $\{T > n\}$ with a variety of other ways to make $X_n(1)$ positive. The ordinary central limit theorem (c.l.t.) reveals that $[X_n(1)]^+ \Rightarrow N^+$, $|X_n(1)| \Rightarrow |N|$, and $[X_n(1)]^2 \Rightarrow N^2$, where N is a normal mean zero, variance one random variable. Next note that $E\{N^+\} = (2\pi)^{-\frac{1}{2}} \cong .399 \leq E\{|N|\} = (2\pi)^{\frac{1}{2}} \cong .798 \leq E\{N^2\} = 1 \leq E\{W^+(1)\} = (\pi/2)^{\frac{1}{2}} \cong 1.253$. Thus we see that on the average conditioning to stay positive throws the random walk out further than does the positive part, absolute value, or square. This is to be expected since conditioning to stay positive serves to eliminate from the probability space all those smallish sample paths which are "flirting with the zero level."

Back to the f.d.d.'s. For $0 < t < 1$ and $x \geq 0$

$$P_n\{X_n(t) \leq x\}$$

$$= r_n^{-1} P\{X_{[nt]}(1) \leq xn^{\frac{1}{2}}/[nt]^{\frac{1}{2}}, T > n\}$$

$$= r_n^{-1} r_{[nt]} \int_0^{xn^{\frac{1}{2}}/[nt]^{\frac{1}{2}}} P_{[nt]}\{X_{[nt]}(1) \in dy\}$$

$$\cdot P\{m_{n-[nt]} > y\sigma[nt]^{\frac{1}{2}}\},$$

where $m_n = \min\{S_r : 0 \leq r \leq n\}$. From here (2.9), the c.l.t. for m_n, and a standard Lebesgue type lemma suffice to show that for $0 < t < 1$

$$(X_n(t) | T > n) \Rightarrow W^+(t) .$$

The idea behind the convergence of the k (>1) dimensional distributions is basically the same.

We use Theorems 15.1 and 15.5 of BILLINGSLEY (1968) to show tightness. For functions $x \in D$ define the modulus of continuity

$$w_x(\delta, a, b) = \sup\{|x(s) - x(t)|\} ,$$

where $0 \leq a < b \leq 1$ and $t-s \leq \delta$. Our task now to complete the proof of $(X_n | T > n) \Rightarrow W^+$ is to show for every $\varepsilon > 0$ that

(2.10) $$\lim_{\delta \to 0} \overline{\lim_{n \to \infty}} P_n\{w_{X_n}(\delta,0,1) \geq \epsilon\} = 0.$$

Note that this will also provide a proof of the existence of W^+ and the fact that $P\{W^+ \in C[0,1]\} = 1$. The first step toward (2.10) is to show that

(2.11) $$\lim_{\tau \to 0} \overline{\lim_{n \to \infty}} r_n^{-1} P\{\sup_{0 \leq s \leq \tau} X_n(s) \geq \epsilon, T > [n\tau]\} = 0.$$

Next decompose the event in (2.10) for $0 < \delta < \tau$, $\tau \in (0,1]$, as

(2.12) $$P_n\{w_{X_n}(\delta,0,1) \geq \epsilon\}$$

$$\leq r_n^{-1} P\{w_{X_n}(\delta,0,1) \geq \epsilon, T > [n\tau]\}$$

$$\leq r_n^{-1} P\{w_{X_n}(\delta,0,1) \geq \epsilon, \sup_{0 \leq s \leq \tau} X_n(x) < \epsilon, T > [n\tau]\}$$

$$+ r_n^{-1} P\{\sup_{0 \leq s \leq \tau} X_n(s) \geq \epsilon, T > [n\tau]\}.$$

The term

(2.13) $$r_n^{-1} P\{w_{X_n}(\delta,0,1) \geq \epsilon, \sup_{0 \leq s \leq \tau} X_n(s) < \epsilon, T > [n\tau]\}$$

$$\leq r_n^{-1} P\{w_{X_n}(\delta, \tau-\delta, 1) \geq \epsilon, \ T > [n\tau]\}$$

$$\leq r_n^{-1} P\{w_{X_n}(\delta, \tau-\delta, 1) \geq \epsilon, \ T > [n(\tau-\delta)]\}$$

$$= r_n^{-1} \ r_{[n(\tau-\delta)]} P\{w_{X_n}(\delta, \tau-\delta, 1) \geq \epsilon\}$$

$$\leq r_n^{-1} \ r_{[n(\tau-\delta)]} P\{w_{X_n}(\delta, 0, 1) \geq \epsilon\}$$

where the equality uses the path structure of the random walk. Since $X_n \Rightarrow W$ and $P\{W \in C[0,1]\} = 1$,

(2.14) $$\lim_{\delta \to 0} \varlimsup_{n \to \infty} P\{w_{X_n}(\delta, 0, 1) \geq \epsilon\} = 0 \ .$$

Now combine (2.12), (2.11), (2.13), and (2.14) to obtain (2.10). This completes a sketch of the proof of

(2.15) THEOREM. <u>If</u> $\mu = 0$, $0 < \sigma^2 < \infty$, $E\{|\xi_1|^3\} < \infty$, <u>and</u> ξ_1 <u>is non-lattice or integer-valued with span 1, then</u> $(X_n | T > n) \Rightarrow W^+$.

The nagging question remains as to whether the finite third moment condition is really necessary. One would hope and expect that it isn't, however, a proof eludes us.

In passing we mention the fact that if $\mu > 0$ and $0 < \sigma^2 < \infty$, then $((S_{[n \cdot]} - \mu n \cdot)/\sigma n^{\frac{1}{2}} | T > n) \Rightarrow W$ as $n \to \infty$. That is, the conditioning on $T > n$ plays no role when the random walk is drifting to $+\infty$. In fact $\{T > n\}$ is the "wrong" event to condition on in this situation. We really should be conditioning on $\{n < T < \infty\}$ to obtain an interesting new result. Our conjecture is that

$(S_{[n \cdot]}/\sigma n^{\frac{1}{2}} | n < T < \infty) \Rightarrow W_0^+$.

We turn our attention now to the random functions $\{(X_n | T = n) : n \geq 1\}$. From (2.2) we see $f_n = P\{T=n\}$ decreases to zero even faster than does r_n: $f_n/r_n \sim (2n)^{-1}$ as $n \to \infty$. Thus we need a more powerful microscope to examine $(X_n | T = n)$ than was required for $(X_n | T > n)$. The key to our analysis in this case is the local behavior of the first passage times of the random walk to a fixed level. To that end define for $x \leq 0$

$$T_x = \inf\{n > 0 : S_n \leq x\}.$$

Before stating the next result we introduce two sets of conditions on the distribution function of ξ_1. Here f is the characteristic function of ξ_1.

(2.16) $E\xi_1 = 0$, $E|\xi_1|^5 < \infty$, and the d.f. of ξ_1 has an absolutely continuous component;

(2.17) $E\xi_1 = 0$, $E|\xi_1|^6 < \infty$, and $\limsup\limits_{|t| \to \infty} |f(t)| < 1$.

Then the following lemma can easily be obtained from NAGAEV (1970), Theorems 1 and 2.

(2.18) LEMMA. <u>Assume either conditions (2.16) or (2.17) hold. Then for</u> $0 < a < b < \infty$ <u>and</u> $0 < \delta < 1$

(2.19) $P\{T_{-y \sigma n^{\frac{1}{2}}} = n - [nt]\} \sim \dfrac{n^{-1} y \exp\{-y^2/2(1-t)\}}{\sqrt{2\pi(1-t)^3}}$

<u>uniformly in</u> $y \in [a,b]$ <u>and</u> $t \in [0,\delta]$ <u>as</u> $n \to \infty$.

After this preparation we can plunge into the proof of convergence of the f.d.d.'s.

(2.20) THEOREM. <u>If conditions (2.16) or (2.17) hold, then for</u> $k \geq 1$ <u>and</u> $0 \leq t_1 < t_2 < \ldots t_k \leq 1$ <u>as</u> $n \to \infty$

(2.21) $(X_n(t_1), \ldots, X_n(t_k) | T = n) \Rightarrow (W_0^+(t_1), \ldots, W_0^+(t_k))$

Proof. The claim (2.21) is trivial for $k = 1$ and $t = 0$. Next consider $k = 1$ and $t = 1$. For $\epsilon > 0$

(2.22) $Q_n\{X_n(1) \leq -\epsilon\} = f_n^{-1} P\{T > n-1, S_n \leq -\epsilon \sigma n^{\frac{1}{2}}\}$

$= f_n^{-1} \int_{(0,\infty)} P\{T > n-1, \dfrac{S_{n-1}}{\sigma n^{\frac{1}{2}}} \in dy\}$

$\cdot P\{\xi_1 \leq -\epsilon \sigma n^{\frac{1}{2}} - y\}$.

Using Chebyshev $P\{\xi_1 \leq -\epsilon \sigma n^{\frac{1}{2}} - y\} \leq E|\xi_1|^3/\sigma^3 \epsilon^3 n^{3/2}$ for all $y > 0$. So from (2.13) we have the estimate

$$Q_n\{X_n(1) \le -\epsilon\} \le E|\xi_1|^3 r_{n-1}/n^{3/2} f_n \sigma^3 \epsilon^3 \to 0$$

as $n \to \infty$ by virtue of Lemmas 2.1 and 2.2. Thus $(X_n(1)|T=n) \Rightarrow 0 = W_0^+(1)$. Now consider $k = 1$, $t \in (0,1)$ and take $0 < a < b$. For all such choices of a, b, and t it will suffice to show that

$$Q_n\{a < X_n(t) \le b\} \to P\{a < W_0^+(t) \le b\} .$$

We can immediately write, using the sample path structure of $\{S_n : n \ge 0\}$,

$$(2.23) \quad Q_n\{a < X_n(t) \le b\}$$

$$= f_n^{-1} r_{[nt]} \int_{(a,b]} P\left\{\frac{S_{[nt]}}{\sigma n^{\frac{1}{2}}} \in dy \,\Big|\, T > [nt]\right\}$$

$$\cdot P\{T_{-y\sigma n^{\frac{1}{2}}} = n - [nt]\}.$$

Next use (2.1), (2.2), (2.15), (2.19) and a version of the continuous mapping theroem (e.g., [4], Theorem 5.5, or [11], Lemma 2.18) to conclude from (2.23) that

$$(2.24) \quad \lim_{n \to \infty} Q_n\{a < X_n(t) \le b\}$$

$$= \frac{2}{\sqrt{2\pi(1-t)^3}} \int_{(at^{-\frac{1}{2}}, bt^{-\frac{1}{2}}]} y \, e^{-y^2/2(1-t)}$$

$$\cdot p(0,0,1,y) \, dy$$

Finally use the fact that $p(0,0,1,y) = y \exp(-y^2/2)$ plus a change of variables to show that the right-hand side of (2.24) is $P\{a < W_0^+(t) \leq b\}$. For $k > 1$ the idea of the proof is the same but now exploits the fact that $(X_n(t_1), \ldots, X_n(t_{k-1})|T > n) \Rightarrow (W^+(t_1), \ldots, W^+(t_{k-1}))$. As the details of the calculation only involve checking some complicated expressions, we shall omit them.

We have only succeeded in showing that $(X_n|T = n) \Rightarrow W_0^+$ under overly restrictive conditions on ξ_1; i.e., ξ_1 is integer-valued and left-continuous. As these results are still incomplete, we shall not bother to indicate a proof. Suffice it to say that the greatest difficulty comes in estimating $P\{T_{-y\sigma n^{\frac{1}{2}}} = n - [nt]\}$ uniformly in y.

3. Conditioned Random Partial Sums and Renewal Processes

BILLINGSLEY (1968), Section 17, has shown us that whenever we have a Donsker-type theorem we can also obtain a corresponding result for random partial sums and renewal processes. Our purpose here is to illustrate that this also holds for conditioned limit theorems; see [11], Section 4 for details.

Suppose $\{N(t):t \geq 0\}$ is a renewal process defined on the same probability space $(\Omega, \mathfrak{F}, P)$, as the random walk. Let the times between renewal epochs be $\{u_i: i \geq 1\}$ with $E\{u_1\} = \lambda^{-1}$, $0 < \lambda < \infty$, and $E\{u_1^2\} < \infty$. It should be emphasized that no independence assumptions are made between the random walk and renewal process. For $n \geq 1$ define the random functions

$$Z_n(t) = S_{N(nt)}/\sigma(\lambda n)^{\frac{1}{2}}, \quad 0 \leq t \leq 1,$$

and the conditioned random functions $\{(Z_n|T>N(n)): n \geq 1\}$. By conditioning on the value of $N(n)$ it is easy to show that $P\{T > N(n)\} \sim c(\lambda n)^{\frac{1}{2}}$ as $n \to \infty$. Similarly by conditioning on the values of $N(nt_1)$, ..., $N(nt_k)$, $N(n)$ and using (2.15) we can show that

$$(Z_n(t_1), \ldots, Z_n(t_k)|T > N(n)) \Rightarrow (W^+(t_1), \ldots, W^+(t_k)).$$

The proof of tightness of $\{(Z_n|T>N(n)): n \geq 1\}$ is basically the same as that employed in showing (2.15). Hence we conclude

(3.1) THEOREM. <u>Under the conditions of</u> (2.15) $(Z_n|T > N(n)) \Rightarrow W^+$ as $n \to \infty$.

Another proof of (3.1) using Billingsley's random change of time idea also seems possible.

In the case of conditioning Z_n on $\{T = n\}$ we would expect a similar result but suspect that one might have to assume $E\{u_1^4\} < \infty$. However, such details should be left until we have established that $(X_n|T = n) \Rightarrow W_0^+$ in some greater generality.

Turning now to conditioned limit theorems for renewal processes assume that $\{\xi_k: k \geq 1\}$ are nonegative with finite mean $\mu > 0$ and otherwise satisfy the conditions of (2.15). Next let

$$Y_n(t) = (S_{[nt]} - \mu nt)/\sigma n^{\frac{1}{2}}, \quad 0 \le t \le 1$$

and $\tilde{\Lambda}_n = \{S_k - \mu k > 0 : 1 \le k \le n\}$. By applying (2.15) to the random variables $\xi_k - \mu$ we can conclude that $(Y_n|\tilde{\Lambda}_n) \Rightarrow W^+$. Denote the renewal process, with rate $\mu^{-1} \equiv \lambda$, associated with the sequence $\{\xi_k : k \ge 1\}$ by $\{N(t) : t \ge 0\}$:

$$N(t) = \#\{n \ge 1 : S_n \le t\}.$$

Form the random functions

$$N_n(t) = (\lambda nt - N(nt))/\sigma\lambda^{3/2} n^{1/2}, \quad 0 \le t \le 1,$$

and let $\Gamma_n = \{\lambda\tau - N(\tau) > 0 : 0 < \tau \le n\}$. Our goal is to show that $(N_n|\Gamma_n) \Rightarrow W^+$. A quick proof of this result follows from the method of VERVAAT (1972), Section 5. Vervaat's method exploits the fact that partial sum and renewal processes are essentially inverses of each other. Here is a sketch of the idea. From (2.15) we know that $((S_{[n\lambda \cdot]} - n \cdot)/\sigma(n\lambda)^{\frac{1}{2}} | \tilde{\Lambda}_{[n\lambda]}) \Rightarrow W^+$ and hence also $((n^{-1}S_{[n\lambda \cdot]} - \cdot)/\sigma\lambda^{\frac{1}{2}}n^{-\frac{1}{2}} | \tilde{\Lambda}_{[n\lambda]}) \Rightarrow W^+$. Now use [26], p. 251, plus the fact that $P\{W^+ \in C\} = 1$ to conclude that $(N_n|\tilde{\Lambda}_{[n\lambda]}) \Rightarrow W^+$. Next observe that $\tilde{\Lambda}_{[n\lambda]} \subset \Gamma_n \subset \tilde{\Lambda}_{[n\lambda]+1}$. From here a simple argument yields

(3.2) THEOREM. <u>If $\{\xi_k : k \geq 1\}$ are non-negative, $E\{\xi_1\} > 0$, and otherwise satisfy the conditions of (2.15), then</u> $(N_n | \Gamma_n) \Rightarrow W^+$ <u>as</u> $n \to \infty$.

Consider now the set $\Pi_u = \{\lambda\tau - N(\tau) : 0 < \tau < u; \lambda u - N(u) \leq 0\}$. Then the same method should yield $(N_u | \Pi_u) \Rightarrow W_0^+$ in those cases for which $(X_n | T = n) \Rightarrow W_0^+$.

4. Random Walks with Negative Drift

Our concern in this section is random walks with $\mu < 0$ conditioned on $T > n$; this case corresponds to a GI/G/1 queue in light traffic, $\rho < 1$, conditioned on the first busy period not expiring by the nth customer. Since the random walk is drifting to $-\infty$ by virtue of the strong law, the probabilities r_n and f_n will converge to zero much faster than is the case when $\mu = 0$. The nature of the results and the method of proof here is completely different. For complete details see [12].

It is generally the case that when a random walk is strongly attracted to the origin any limit results depend on the full distribution of ξ_1, not just first and second moments. Furthermore, it is often necessary to assume that ξ_1 has an exponential upper tail. In particular, we shall assume throughout this section that the following conditions hold:

(4.1) $\qquad -\infty \leq \mu < 0;$

(4.2) $\theta(s) = E\{\exp(s\xi_1)\}$ converges for real $s \in [0,a)$ for some $a > 0$;

(4.3) $\theta(s)$ attains its infimum at a point τ, $0 < \tau < a$, where $\theta(\tau) \equiv \gamma < 1$ and $\theta'(\tau) = 0$; and

(4.4) if ξ_1 is lattice, then $P\{\xi_1 = 0\} > 0$.

The chief virtue of these assumptions is that they imply that as $n \to \infty$

(4.5) $$P\{S_n > 0\} \sim \frac{\gamma^n}{(2\pi n)^{\frac{1}{2}}} \cdot \frac{1}{\alpha\tau}$$

and for $u \geq 0$

(4.6) $$E\{e^{-uS_n}:S_n > 0\} \sim \frac{\gamma^n}{(2\pi n)^{\frac{1}{2}}} \cdot \frac{1}{\alpha(\tau+u)},$$

where $\alpha^2 = \theta''(\tau)/\gamma$, $0 < \alpha < \infty$; see BAHADUR and RAO (1960), Theorem 1, for (4.5). The same proof yields (4.6).

Our goal here is to show that when (4.1)-(4.4) hold then $(S_n|T > n) \Rightarrow S^*(\mu)$ where $S^*(\mu)$ is a complicated non-degenerate random variable whose Laplace transform we shall display; DALEY (1968) obtained this result for integer-valued, left-continuous random walks. First note that this is an ordinary c.l.t. and not a functional c.l.t. as we had in Sections 2 and 3; we know of no f.c.l.t. in this case. The next point to observe is that we need not normalize S_n by $n^{\frac{1}{2}}$ to get convergence to a non-degenerate limit. This is basically because the negative drift does not permit the random walk to become very large.

The starting point for this analysis is (2.7). If we set

$$f_n(u) = E\{e^{-uS_n}|T > n\}$$

and

$$\varphi_n(u) = E\{e^{-uM_n}\}$$

for $u \geq 0$, then (2.7) yields the relation

$$(4.7) \quad f_n(u) = 1 - \sum_{k=0}^{n-1} (r_k/r_n)[\varphi_{n-k-1}(u) - \varphi_{n-k}(u)].$$

From random walk theory we know that

$$(4.8) \quad R(s) = \sum_{n=0}^{\infty} r_n s^n = \exp\left\{\sum_{n=1}^{\infty} \frac{s^n}{n} P\{S_n > 0\}\right\}$$

and

$$(4.9) \quad \Phi(s) = \sum_{n=0}^{\infty} \varphi_n(u) s^n = \exp\left\{\sum_{n=1}^{\infty} \frac{s^n}{n} E\{e^{-uS_n^+}\}\right\}$$

If we let $F(s) = \sum_{n=0}^{\infty} r_n f_n(u) s^n$, then using (4.7) it is easy to show that

$$(4.10) \quad F(s) = (1 - s) R(s) \Phi(s).$$

To study the behavior of two sequences $\{r_n : n \geq 0\}$

and $\{f_n(u) : n \geq 0\}$ we use the following two lemmas; see [12] for proofs.

(4.11) LEMMA. Let $\sum_{n=0}^{\infty} d_n s^n = \exp\{\sum_{n=1}^{\infty} b_n s^n\}$ for $|s| \leq 1$. If $b_n \geq 0$ and $b_n = O(n^{-3/2})$, then $d_n = O(n^{-3/2})$ as $n \to \infty$.

(4.12) LEMMA. Let $c_n, d_n \geq 0$, $c_n \sim cn^{-\frac{1}{2}}$ with $c > 0$, $\sum_{n=0}^{\infty} d_n = d < \infty$, and $d_n = O(n^{-1})$. If $a_n = \sum_{j=0}^{n-1} c_{n-j} d_j$, then $a_n \sim cdn^{-\frac{1}{2}}$ as $n \to \infty$.

With these lemmas in hand we proceed to

(4.13) THEOREM. If conditions (4.1)-(4.4) hold, then as $n \to \infty$

$$r_n \sim \frac{\gamma^n}{n^{3/2}} \cdot \frac{1}{(2\pi)^{\frac{1}{2}} \alpha \tau} \cdot \exp\{\sum_{n=1}^{\infty} \frac{\gamma^{-n}}{n} P\{S_n > 0\}\}.$$

Proof. Differentiate (4.8) with respect to s and obtain

(4.14) $\sum_{n=1}^{\infty} n r_n \gamma^{-n} s^n = \exp\{\sum_{n=1}^{\infty} \frac{\gamma^{-n}}{n} P\{S_n > 0\} s^n\}$
$\sum_{n=1}^{\infty} \gamma^{-n} P\{S_n > 0\} s^n$,

where $|s| \leq 1$. Now set in the notation of (4.11) and (4.12), $a_n = nr_n\gamma^{-n}$, $c_n = \gamma^{-n}P\{S_n > 0\}$, $b_n = n^{-1}c_n$, and

$$\sum_{n=0}^{\infty} d_n s^n = \exp\left\{\sum_{n=1}^{\infty} \frac{\gamma^{-n}}{n} P\{S_n > 0\}s^n\right\}.$$

From (4.14) we see that $a_n = \sum_{j=0}^{n-1} c_{n-j}d_j$. Next use (4.5) to see that $b_n = O(n^{-3/2})$ and (4.11) to see that $d_n = O(n^{-3/2})$. Thus $\sum_{n=0}^{\infty} d_n = d < \infty$. We have now checked that all the conditions of (4.12) hold so that $a_n \sim cdn^{-\frac{1}{2}}$ which is the desired result.

(4.15) THEOREM. *If conditions* (4.1)-(4.4) *hold, then as* $n \to \infty$

$$r_n f_n(u) \sim \frac{\gamma^n}{n^{3/2}} \cdot \frac{1}{(2\pi)^{\frac{1}{2}}\alpha(\tau+u)} \cdot \exp\left\{\sum_{n=1}^{\infty} \frac{\gamma^{-n}}{n} \cdot E\{e^{-uS_n}:S_n > 0\}\right\}.$$

Proof. From (4.8), (4.9), and (4.10) we obtain

$$(4.16) \sum_{n=0}^{\infty} r_n f_n(u)\gamma^{-n}s^n = \exp\left\{\sum_{n=1}^{\infty} \frac{\gamma^{-n}}{n} E\{e^{-uS_n}:S_n>0\}s^n\right\},$$

$$|s| \leq 1,$$

where we have used the fact that $(1-s) =$

$\exp\left\{-\sum_{n=1}^{\infty}\frac{s^n}{n}\right\}$, and that $E\left\{e^{-uS_n}:S_n>0\right\}=$
$E\left\{e^{-uS_n^+}\right\}+P\{S_n>0\}-1$. Next differentiate (4.16) with respect to s and use (4.6) together with the method employed in (4.13) to obtain the result.

Combining (4.13) and (4.15) we have our main result.

(4.17) THEOREM. <u>If conditions (4.1)-(4.4) hold, then</u>

$$\lim_{n\to\infty} f_n(u) = \frac{\tau}{\tau+u} \cdot \exp\left\{\sum_{n=1}^{\infty}\frac{\gamma^{-n}}{n}\left[E\{e^{-uS_n^+}\}-1\right]\right\} \equiv f(u).$$

Thus we have shown that $(S_n|T>n) \Rightarrow S^*(\mu)$, where $S^*(\mu)$ has Laplace transform $f(u)$. Note that $S^*(\mu)$ can be represented as the sum of two independent random variables, one of which is exponential with parameter τ and the other is closely related to $M = \sup\{S_r : r \geq 0\}$. Then $E\{e^{-uM}\}$

$$= \exp\left\{\sum_{n=1}^{\infty}\frac{1}{n}\left[E\{e^{-uS_n^+}\}-1\right]\right\}.$$

While $f(u)$ can be explicitly evaluated in some simple cases, for example, the M/M/1 queue, it is in general untractable. The same thing can be said for the $E\{e^{-uM}\}$. In queueing theory the random variable M has the same distribution as the stationary waiting time $W(\mu)$; here $W_n \Rightarrow W(\mu)$ as $n \to \infty$. KINGMAN (1962) in an attempt to obtain a simple approximation for $W(\mu)$ considered the heavy traffic situation in which one has a family of queueing systems with $\mu \uparrow 0$ ($\rho \uparrow 1$). He found under some mild regularity assumptions that $(2|\mu|/\sigma^2)W(\mu) \Rightarrow \Gamma(1)$ as $\mu \uparrow 0$, where $\Gamma(1)$ is an exponential random variable with density e^{-x}, $x \geq 0$. In our case of the conditioned limit of W_n a similar result holds. Again under some mild regularity conditions we find that as $|\mu| \uparrow 0$

(4.18) $\qquad (|\mu|/\sigma^2)S^*(\mu) \Rightarrow \Gamma(2),$

where $\Gamma(2)$ has density xe^{-x} for $x \geq 0$, see [12], Section 3 for the proof. Thus (4.18) provides an approximation for the unwieldly $S^*(\mu)$ when μ is negative but close to zero.

5. Open Problems

It is our feeling that work on conditioned limit theorems has only just begun and that there are many interesting problems to be solved. In this section we mention a few of these.

(5.1) Show under some reasonable conditions, such as those of (2.15), that $(X_n | T_n = n) \Rightarrow W_0^+$.

(5.2) Show under appropriate conditions that when $\mu \neq 0$, $((S_{[n \cdot]}/\sigma n^{\frac{1}{2}} | n < T < \infty) \Rightarrow W_0^+$

(5.3) Consider hitting times of the form $T_n(x) = \inf\{n > 0 : S_n \notin [0, xn^{\frac{1}{2}}]\}$ for $x > 0$. Obtain a limit law for $(X_n | T_n(x) > n)$.

(5.4) Obtain local limit theorems for $(S_n|T > n)$ and $(S_n|T = n)$ in the spirit of KAIGH (1974).

(5.5) The principal reason for obtaining f.c.l.t's is to easily deduce limit theorems for functionals of the processes in question. For example, in the case of (2.15) $h[(X_n|T > n] \Rightarrow h(W^+)$ for appropriate functions h. Results of this type are of interest only if we know the distribution of, for example, $h(W^+)$. At this point we should use the idea of invariance principles to find the distribution of functions of W^+ and W_0^+.

REFERENCES

[1] BAHADUR, R. R. and RAO, R. R. (1960). On deviations of the sample mean. Ann. Math. Statist. 31, 1015-1027.

[2] BELKIN, B. (1970). A limit theorem for conditioned recurrent random walk attracted to a stable law. Ann. Math. Stat. 41, 146-163.

[3] BELKIN, B. (1972). An invariance principle for conditioned random walk attracted to a stable law. Z. Wahrscheinlichkeitstheorie verw. Geb. 21, 45-64.

[4] BILLINGSLEY, P. (1968). Convergence of Probability Measures. John Wiley, New York.

[5] BOROVKOV, A. A. (1970). Factorization identities and properties of the distribution of the supremum of sequential sums. Theory Probability Appl. 15, 359-402 (English translation).

[6] DALEY, D. (1969). Quasi-stationary behaviour of a left-continuous random walk. Ann. Math. Stat. 40, 532-539.

[7] DONSKER, M. (1951). An invariance principle for certain probability limit theorems. Mem. Amer. Math. Soc. **6**.

[8] DWASS, M. and KARLIN, S. (1963). Conditioned limit theorems. Ann. Math. Stat. 34, 1147-1167.

[9] FELLER, W. (1971). An Introduction to Probability Theory and Its Applications, Volume 2. John Wiley and Sons, New York.

[10] IGLEHART, D. L. (1971). Functional limit theorems for the queue GI/G/I in light traffic. Advances in Appl. Probability 3, 269-281.

[11] IGLEHART, D. L. (1974a). Functional central limit theorems for random walks conditioned to stay positive. Ann. Probability 2, 608-619.

[12] IGLEHART, D. L. (1974b). Random walks with negative drift conditioned to stay positive. To appear in J. Appl. Probability, 1974.

[13] ITÔ, K. and McKEAN, H. P., Jr. (1965). Diffusion Processes and Their Sample Paths. Springer-Verlag, Berlin.

[14] KAIGH, W. D. (1974). A conditional local limit theorem and its application to random walk. Bull. Amer. Math. Soc. 80, 769-770.

[15] KINGMAN, J. F. C. (1962). On queues in heavy traffic. J. Roy Statist. Soc., Ser. B 25 383-392.

[16] LIGGETT, T. M. (1968). An invariance principle for conditioned sums of independent random variables. J. Math. Mech. 18, 559-570.

[17] LIGGETT, T. M. (1970a). Weak convergence of conditioned sums of independent random vectors. Trans. Amer. Math. Soc. 152, 195-213.

[18] LIGGETT, T. M. (1970b). Convergence of sums of random variables conditioned on a future change of sign. Ann. Math. Stat. 41, 1978-1982.

[19] McLEISH, D. L. (1974). Dependent central limit theorems and invariance principles. Ann. Probability 2, 620-628.

[20] NAGAEV, S. V. (1969). An estimate for the speed of convergence of the distribution of maximum sums of independent random variables. Siberian Math. J. 10, 443-458 (English translation).

[21] NAGAEV, S. V. (1970). Asymptotic expansions for the distribution function of the maximum of a sum independent identically distributed random quantities. Siberian Math. J. 11, 288-309.

[22] PORT, S. C. and STONE, C. J. (1971). Infinitely divisible processes and their potential theory, Parts I, II. Ann. Inst. Fourier, Grenoble, 21(2), 157-275; 21 (4), 179-265.

[23] PROHOROV, Yu. V. (1956). Convergence of random processes and limit theorems in probability theory. Theor. Probability Appl. 1, 157-214 (English translation).

[24] SKOROHOD, A. V. (1956). Limit theorems for stochastic processes. Theor. Probability Appl. 1, 262-290 (English translation).

[25] SPITZER, F. (1960). A Tauberian theorem and its probability interpretation. Trans. Amer. Math. Soc. 94, 150-169.

[26] VERVAAT, W. (1972). Functional central limit theorems for processes with positive drift and their inverses. Z. Wahrscheinlichkeitstheorie verw. Geb. 23, 245-253.

Canonical Representations Of Equivalent Gaussian Processes*

by G. Kallianpur
University of Minnesota

Introduction. In the theory of Gaussian processes depending on a continuous parameter, there are two concepts, apparently originating independently of each other, which have been investigated fairly extensively in recent years. The first one (the older of the two) is that of the canonical representation, introduced by Lévy and later developed by Cramér and Hida [1,2,6,15]. The second, that of the non-anticipative or causal representation arose out of the study of Gaussian processes which are equivalent to each other, i.e., whose measures are mutually absolutely continuous. The existence of such representations in general was shown by Kallianpur and Oodaira [12] and by Kailath and Duttweiler [10],

---

* Presented at the Summer Research Institute on Statistical Inference for Stochastic Processes held at Indiana University, Bloomington, Indiana, July 31-August 9, 1974. This research was supported in part by National Science Foundation Grant GP 30694 X.

while its existence in a special but important case was discovered (by a different method) earlier by Hitsuda [7].

The connection between the two types of representations for the class of equivalent Gaussian processes forms the topic of the present article. Suppose $X = (X(t))$ is a mean continuous Gaussian process having a canonical representation. If $Y = (Y(t))$ is any Gaussian process equivalent to X it is shown that Y also has a canonical representation with the same multiplicity and sequence of spectral types as that of X. Furthermore, the latter is given by the non-anticipative representation of Y with respect to X.

Given the canonical representation of X the problem of deriving that of Y is discussed and as an important example, the kernel of the representation of Y is explicitly obtained for the case of arbitrary continuous multiplicity. The latter problem has also been solved very recently by Hitsuda using martingale methods [9]. The case of multiplicity one has been treated in [13]. Also in [13] the general non-anticipative representation has been obtained in terms of operator integrals, an approach different from the one taken here.

1. <u>Canonical representations</u>. Let $X = (X(t))_{t \in T}$ be a Gaussian stochastic process assumed to be mean continuous and having zero mean and covariance function Γ_X. The index set T is either the real line or the closed interval [0,1]. The underlying probability space $(\Omega, \underline{A}, P)$ on which X is defined is not of particular significance for the present. Denote by

$L(X)$ the closed linear subspace of the Hilbert space $L^2(\Omega,\underline{A},P)$ spanned by the family $\{X(t),\ t \in T\}$. For each $t \in T$, $L(X;t)$ will similarly denote the subspace generated by $\{X(s),\ s \leq t,\ s \in T\}$. Throughout this section we shall make the assumption

(1.1) $\bigcap_{t \in T} L(X;t) = 0.$

If $T = [0,1]$, (1.1) reduces to $L(X;0) = 0$. The condition (1.1) which, in the theory of prediction is known as a property which makes X "completely non-deterministic", can be dispensed with here but is retained for convenience.

We introduce next the family of orthoprojectors $P^X(t)$ on $L(X)$ with range $L(X;t)$. Then it is easy to see that $P^X(t)$ $(t \in T)$ is a spectral family or a resolution of the identity with $P^X(t-0) = P^X(t)$ for each t (or each $t > 0$ if T is $[0,1]$). The last property is a consequence of the fact that since X is a quadratic mean continuous process, $X(t-0)$, $X(t+0)$ exist (in quadratic mean) and $X(t-0) = X(t)$. A theory of representation of Gaussian processes which was initiated by P. Lévy [15] has been developed by H. Cramér [1] and T. Hida [6]. The work of the latter authors is directly based on multiplicity theory in separable Hilbert spaces and the theorm of Hellinger-Hahn which is treated in detail in Chapter 7 of M. H. Stone's book [21]. We shall content ourselves with introducing the minimal amount of terminology and notation necessary for the definition of a representation. For any element f in L(X) let ρ_f be the finite measure on the Borel sets of the real line (or of [0,1] if T = [0,1]) given by $\rho_f(\Delta) = \|P^X(\Delta)f\|^2$. The family of all finite Borel measures on the line (whose measure functions are taken to be left-continuous) is divided into equivalence classes by the relation of equivalence (i.e. mutual absolute continuity) of measures. If ρ denotes the equivalence class to which ρ_f belongs, ρ will be called the <u>spectral type</u> of f. If ρ and σ are two types we write $\rho > \sigma$ to mean that if σ_f belongs to σ and ρ_g belongs to ρ then $\sigma_f << \rho_g$. It is clear that a partial ordering is thus introduced in the family of all types. The following result which is deduced from the theory in [21] asserts the existence of a representation for X(t) regarded as an element of L(X;t). For convenience of application we take T = [0,1] below.

<u>Theorem A</u>. (Cramér [1]). Let X be the Gaussian process defined above and satisfying condition (1.1). Then there exists a sequence (f_i) (i=1,...,N) of elements in L(X) with the following properties. Define

$\xi_i(t) = P^X(t)f_i$ and let ρ_i be the spectral type of ρ_{f_i} (which we shall also denote by ρ_i).

(1.2) $\rho_1 > \rho_2 > \ldots > \rho_N$

(1.3) The $\xi_i(t)$ are mutually independent Gaussian additive processes with variance functions $\rho_i(t)$ and with zero means.

(1.4) $\quad L(\xi_i;t) \perp L(\xi_j;t) \quad (i \neq j)$

(These Hilbert spaces are defined in the same way as $L(X;t)$).

(1.5) $\quad L(X;t) = \sum_{i=1}^{N} \oplus L(\xi_i;t)$

(1.6) $\quad X(t) = \sum_{i=1}^{N} \int_0^t F_i(t,u) \, d\xi_i(u)$

where $F_i(t,u) = 0$ if $u > t$ and

(1.7) $\quad \sum_{i=1}^{N} \int_0^1 F_i^2(t,u) \, d\rho_i(u) < \infty$.

Furthermore, the number N (which may be infinite) is called the multiplicity of $X(t)$ and is uniquely determined as the number of elements in the sequence (1.2). The sequence of spectral types (1.2) is also uniquely determined although the choice of $\{f_i\}$ is not. No representation of the form (1.6) with these properties exists for any smaller value of N.

We shall call (1.6) a <u>canonical representation</u> of $X(t)$ (generalized canonical representation in the terminology of Hida [6]). It can be seen that (1.6) provides a version of the Gaussian process X on some probability space. For suppose $(\Omega, \underline{A}, P)$ is some model on which are defined mutually independent Gaussian additive processes $\xi_i(t,\omega)$, $(t \in T, \omega \in \Omega)$ such that (suppressing ω as usual). $E\xi_i(t) = 0$ for all t and $E\xi_i(t) = \rho_i(t)$ where ρ_i is the ρ_{f_i} of the preceding theorem. On Ω we may now define the stochastic integral $Z_i(t,\omega) = \int_0^t F_i(t,u) \, d\xi_i(u,\omega)$ so as to make Z_i a (t,ω)-measurable process (Doob [3],

p. 430). (Note that in (1.6), the $F_i(t,u)$ may be assumed to be measurable in (t,u).). The set Λ of (t,ω) values for which $\sum_{i=1}^{N} Z_i(t,\omega)$ converges is measurable and from (1.6) we have $P(\Lambda_t) = 1$ for each t, $\Lambda_t = \{\omega \in \Omega: (t,\omega) \in \Lambda\}$. If we now define $Z(t,\omega)$ to be equal to $\sum_{i=1}^{N} Z_i(t,\omega)$ for (t,ω) in Λ and zero (say) on Λ^c it follows that $Z = \{Z(t,\omega)\}$ is a version of X on $(\Omega,\underline{\underline{A}},P)$. Z may be referred to as a stochastic process which is a representation of the stochastic process X. The functions $\{F_i\}$ are called the kernel of the particular representation Z or of (1.6). We shall often use the same letter X to denote Z when there is no possibility of confusion.

Some clarificatory remarks seem desirable before proceeding further. The canonical representation of X is essentially unique in the following sense. Suppose $X' = (X'(t))$ be another version of defined possibly not on the same probability space as X. Let $L(X')$ be the corresponding Hilbert space. Since X' and X have the same covariance function the map $U: X(t) \rightarrow X'(t)$ and defined in the obvious way to be linear extends to a unitary transformation from $L(X)$ onto $L(X')$. Furthermore, we have $U \, P^X(t) \, U^{-1} = P^{X'}(t)$ for each t. Hence condition (1.1) is satisfied for X' and it follows from the theorem above that X' has a canonical representation of the form (1.6) with the same kernel $\{F_i(t,u)\}$, the same multiplicity N and with $\xi_i(t)$ replaced by

$\xi_i'(t)$ in (1.6) satisfying (1.3), (1.4) and with the same spectral functions $\rho_i(t)$. Thus, in particular, all versions of the Gaussian process X have the same canonical representation which in addition, can be chosen to be a measurable process given by $Z(t,\omega)$ on a suitable probability space.

2. **Canonical representations of equivalent Gaussian processes.** Consider a second Gaussian process $Y = Y(t)$, mean continuous with zero mean and covariance function Γ_y. Suppose that (1.1) holds for Y. According to Theorem A, Y has a canonical representation. We shall say that X and Y have equivalent canonical representations if their canonical representations have the same spectral types and the same multiplicity. It is a well-known fact that multiplicity and spectral types are unitary invariants, (Stone [21]). Hence X and Y have (unitarily) equivalent canonical representations if and only if there exists a unitary transformation from $L(X)$ onto $L(Y)$ such that

(2.1) $\quad U\ P^x(t)\ U^{-1} = P^y(t)$

for every t. Suppose (2.1) holds. If $\{f_i\}$ are the elements appearing in Theorem A for the representation of X, $\{g_i\}$ where $g_i = Uf_i$ may be chosen for the representation of Y and $\rho_{g_i} = \rho_{f_i}$.

Since we have assumed the means to be zero, the covariance Γ_x and Γ_y of the Gaussian processes X and Y determine probability measures P and Q on $(\mathcal{R}^T, \underline{B}(\mathcal{R}^T))$. Often, it is convenient to put the measures more in evidence. Let P and Q be complete probability measures on a measurable space $(\Omega,\underline{A})$ on which is defined a family of real random variables $X(t,\omega)$ ($t \in T$). Suppose that $\{X(t)\}$ is a mean continuous, zero mean Gaussian process with covariance function Γ_x under P and Γ_y under Q. The two Gaussian processes under consideration are $\{X(t), P\}$, and $\{X(t), Q\}$. They are equivalent if the measures P and Q are equivalent when restricted to the relevant σ-field, i.e., if $P \equiv Q$ relative to $\underline{B}(X)$ where $\underline{B}(X) = \sigma[X(t), t \in T]$. Except when we are interested in obtaining versions possessing special properties (such as measurability, continuity of sample paths, etc.) we shall continue to designate the processes as X and Y. (It would have been more appropriate, perhaps, to use the terminology Γ_x-process and Γ_y-process since the means -- here taken to be zero -- and the covariance functions uniquely determine the Gaussian distributions under study). The remarks made so far may be summarized in these two statements: (a) The existence of a canonical representation for X depends solely on its covariance function Γ_x which also uniquely determines the multiplicity and the sequence of spectral types of the representation; (b) Whether or not two Gaussian processes X and Y possess (unitarily) equivalent canonical representations is also a property which can be characterized solely in terms of Γ_x and Γ_y.

The investigation of problem (a), of finding usable criteria in terms of Γ_x which help to determine the multiplicity or the spectral types seems, in general, to be difficult. Partial answers have been provided in the recent papers of Hitsuda [8] and of Siraia [20]. In this section we shall consider the qualitative problem (b) in relation to Gaussian measures which are equivalent. In recent years necessary and sufficient conditions for the equivalence of two Gaussian measures have been given by several authors (e.g., see [4,11,16,19]). The result which we find convenient to use is taken from [11] and is stated in terms of the reproducing kernel Hilbert spaces (rkhs) of the covariances. Denote by $H(\Gamma_x)$ the rkhs of Γ_x i.e. the Hilbert space of functions f on T to which belong all the functions $\Gamma_x(\cdot,t)$ ($t \in T$) and such that $f(t) = \langle f, \Gamma_x(\cdot,t) \rangle$.

Theorem B [11]. The Gaussian processes X and Y are equivalent if and only if Γ_y defines a linear operator S on $H(\Gamma_x)$ with the following properties:

(2.2) $\Gamma_y(\cdot,t) = S \Gamma_x(\cdot,t)$ for each $t \in T$,

(2.3) S is a bounded self-adjoint, positive operator,

(2.4) $T = I - S$, is Hilbert-Schmidt,

(2.5) $1 \notin \sigma(T)$, the spectrum of T.

There is a canonical isometry between L(X) and $H(\Gamma_x)$ defined in the obvious way by the relation $X(t) \to \Gamma_x(\cdot,t)$. Operators on L(X) and $H(\Gamma_x)$ which correspond to each other under this isometry will be denoted by the same letter. Before we proceed to the result of this section, it will be necessary to introduce the ideas and terminology used in its proof. Although the reader will find a detailed discussion of these pre-requisites in the paper of Kallianpur and Oodaira [12] we recapitulate them briefly to make the presentation as selfcontained as possible.

Let $T = [0,1]$ and write $P(t)$ in place of $P^X(t)$, the orthoprojector defined on $L(X)$ with range $L(X;t)$. We recall that condition (1.1) holds. Let $L(X;t+) = \bigcap_{s>t} L(X;s)$, and $L(X;t-)$ be the smallest closed linear space containing $L(X;s)$ for $s < t$. Obviously, $0 = L(X;0) \subseteq L(X;0+)$ and $L(X;t-) \subseteq L(X;t) \subseteq L(X;t+)$ for $t > 0$. It is easy to verify that $L(X;t-) = L(X;t)$. Since $L(X)$ is separable the set of discontinuities $E = \{t \in T \mid L(X;t) \neq L(X;t+)\}$ is countable. Let $P(t_j+)$ be the orthoprojector with range $L(X;t_j+)$ for $t_j \in E$. If the space $[P(t_j+) - P(t_j)] L(X)$ has dimension $n_j > 1$, we write it as the sum $\sum_{i=1}^{n_j} \oplus L(j,i)$ of one dimensional subspaces $L(j,i)$. Let $Q(j,k)$ be the orthoprojector with range $\sum_{i=1}^{k} \oplus L(j,i)$. Then the family of orthoprojectors Π consisting of $\{P(t), 0 \le t \le 1\}$, $\{P(t_j+), t_j \in E\}$ and $\{P(t_j) + Q(j,k), k=1,\ldots,n_j-1, t_j \in E\}$ is a maximal chain as defined in the book of Gohberg and Krein [5], i.e. it is ordered by inclusion of its range spaces, contains 0 and I and its gaps (if any) are one-dimensional. (Note that E may be empty.

Also note that the dimension n_j may be infinite in which case we consider $Q(j,k)$ when k ranges from 1 to infinity.) The importance of Π lies in the fact that we can invoke a result of Gohberg and Krein which factorizes the operator S or S^{-1} of Theorem B along the chain Π. In the theorem to be proved below we shall consider only quadratic mean continuous Gaussian processes with zero mean function: the latter assumption is made only for convenience while the mean continuity is a decent guarantee of the separability of the Hilbert spaces we have to work with.

Theorem 1. Let $X = (X(t))$ $(t \in T)$ be a Gaussian process with covariance function Γ_X and satisfying condition (1.1). If $Y = (Y(t))$ is any Gaussian process which is equivalent to X then it has a canonical representation with the same multiplicity and sequence of spectral types as the canonical representation of X.

Proof: By Theorem B, the covariance Γ_Y of Y defines on $H(\Gamma_X)$ the operator S which satisfies conditions (2.2)-(2.5). In particular, $S = I - T$ is invertible and T is a Hilbert-Schmidt operator. Hence by a well-known result due to Gohberg and Krein ([5], Theorem 2.2 of [12]) S has a factorization with respect to the maximal chain Π of the form

(2.6) $\quad S = (I + W^*)D^{-1}(I + W)$

where the operators on the right hand side are described as follows.

(2.7) W is a Hilbert-Schmidt, Volterra operator with Π as an eigenchain, i.e., such that $PWP = WP$ for every $P \in \Pi$. (A Volterra operator is a compact operator whose spectral radius is zero). W^* denotes the transpose of W and is therefore also a Hilbert-Schmidt, Volterra operator but with $\Pi^\perp$ as an eigenchain, where $\Pi^\perp = \{I-P, P \in \Pi\}$.

The operator D is given by the formula

(2.8) $\quad D = I + \sum_j (P_j^+ - P_j^-)[(I - P_j^+ T P_j^+)^{-1} - I](P_j^+ - P_j^-)$,

where $P_j^+ = P(t_j+)$, $P_j^- = P(t_j-) = P(t_j)$ and the summation extends over all t_j belonging to the set E. Furthermore, D has these properties.

(2.9) D is self-adjoint, positive and invertible,

(2.10) I-D is Hilbert-Schmidt, and

(2.11) $DP = PD$ for every $P \in \Pi$.

Setting $\Delta = D^{-1}$ we see that $\Delta^{1/2}$ commutes with all the orthoprojectors P in Π. Write

(2.12) $\quad F = \Delta^{1/2}(I + W)$.

Then from (2.6) $S = F^*F$. Let us now consider the operator on L(X) which corresponds to F under the canonical isometry between L(X) and $H(\Gamma_X)$ and denote it also by F. Define

(2.13) $\quad \eta(t) = FX(t)$, $\quad t \in T$.

Then with $(\,,\,)_{L(X)}$ standing for the inner product in $L(X)$ it follows from (2.2) that

$$(2.14) \quad (\eta(t), \eta(s))_{L(X)} = \langle F\, \Gamma_X(\cdot,t), F\, \Gamma_X(\cdot,s)\rangle$$

$$= \langle S\, \Gamma_X(\cdot,t), \Gamma_X(\cdot,s)\rangle$$

$$= \langle \Gamma_Y(\cdot,t), \Gamma_X(\cdot,s)\rangle$$

$$= \Gamma_Y(t,s)\,.$$

Hence

$$(2.14) \quad (\eta(t), \eta(s))_{L(X)} = (Y(t), Y(s))_{L(Y)},$$

where $\{Y(t)\}$ are elements of the Hilbert space $L(Y)$ of the Y-process. Next from (2.7) and (2.11) we have for each t,

$$\eta(t) = \Delta^{1/2}(I + W_+)X(t)$$
$$= \Delta^{1/2}(I + W_+)\mathbf{P}(t)X(t)$$
$$= \Delta^{1/2}P(t)(I + W_+)P(t)X(t)$$
$$= P(t)\,\Delta^{1/2}(I + W_+)X(t)$$
$$= P(t)\eta(t),$$

so that $\eta(t) \in L(X;t)$. Furthermore, the same reasoning shows $F\, L(X;t) \subseteq L(X;t)$ and so we have $L(\eta;t) \subseteq L(X;t)$. We now show that$_*$

$$(2.15) \quad L(\eta;t) = L(X;t),$$

for every t. Suppose $\zeta \in L(X;t)$, $\zeta \perp L(\eta;t)$ for some t. Then for all s in $[0,t]$, $(F^*\zeta, X(s))_{L(X)} = (\zeta, \eta(s))_{L(X)} = 0$. Hence $(P(t)F^*\zeta, X(s))_{L(X)} = 0$ which implies that $P(t)F^*\zeta = 0$, or

(2.16) $F^*\zeta \in L^{\perp}(X;t)$,

the orthocomplement of $L(X;t)$ in $L(X)$. Since $L(X;t)$ is an invariant subspace of F, $L^{\perp}(X;t)$ is an invariant subspace of F^*. Also $(F^*)^{-1} = D^{1/2}(I+V)$ where $V = (I+W^*)^{-1} - I$. We have (from the definition of V) $V + W^* + W^*V = 0$. Since W^* and $I+V$ are permutable the inequality $r_{W^* + W^*V} \leq r_{W^*} r_{I+V}$ holds ([18], p. 426) where r_A denotes the spectral radius of A. But $r_{W^*} = 0$, so $r_{W^* + W^*V} = 0$, i.e. $W^* + W^*V$ is a Volterra operator since it is obviously compact. Thus V itself is a Volterra operator. Furthermore,

$V = \sum_{n=1}^{\infty} (-1)^n (W^*)^n$, the right hand side converging

in norm. It is readily verified that $L^{\perp}(X;t)$ is an invariant subspace of $(W^*)^n$ for every n, so that the same is true of V. Finally, recalling property (2.11) of D we conclude that $L^{\perp}(X;t)$ is an invariant subspace of $(F^*)^{-1}$. Hence $\zeta = (F^*)^{-1} F^* \zeta \in L^{\perp}(X;t)$ from (2.16). But $\zeta \in L(X;t)$ by assumption, so we must have $\zeta = 0$ and (2.15) is proved. (The process η is called the nonanticipative representation of X).

Let us define

(2.17) $U\eta(t) = Y(t)$

and extend it to be linear on the manifold of finite linear combinations of $\{\eta(t)\}$. (2.14) and (2.15) then show that U extends as a unitary transformation from $L(X)$ onto $L(Y)$ with the properties

(2.18) $\quad UL(X;t) = L(Y;t)$,

(2.19) $\quad UL^{\perp}(X;t) = L^{\perp}(Y;t)$.

(Corresponding statements hold for U^{-1}). If $z \in L(Y)$ is arbitrary we have from (2.18) and (2.19) $U^{-1}z = U^{-1}P^y(t)z + U^{-1}P_{L^{\perp}(Y;t)}z$, implying $P^x(t)U^{-1}z = U^{-1}P^y(t)z$, or $UP^x(t)U^{-1}z = P^y(t)z$. We have thus shown that $UP^x(t)U^{-1} = P^y(t)$ for all t. This is precisely (2.1) and hence proves that Y has a canonical representation which is unitarily equivalent to the canonical representation of X. The theorem is thus proved.

It will be seen from the proof of Theorem 1 that the operator $\Delta^{1/2}(I+W)$ plays a crucial role in determining the canonical representation of a Gaussian process Y which is equivalent to X. In ([13] Theorem 2.1) an expansion of $\Delta^{1/2}(I+W)X(t)$ was given in terms of operator integrals of the kind introduced in [5], and from it was derived the canonical representation of all Gaussian processes equivalent to X in the case when the representation of X has no discrete multiplicity. In this derivation ([13], Theorem 2.2), a detailed proof of which is not given, the final expression given for $Y(t)$ is not correct, due to an oversight in the calculation. (See also [14]). This was kindly pointed out to me by Professor M. Hitsuda who has obtained the result by martingale methods. While Theorem 2.1 of [13] is very general, the passage from operator integrals to a formula involving stochastic integrals does not seem to be easy and moreover, the latter formula can be obtained directly from Theorem 1. We shall first derive the general form of the canonical representation of Y in Theorem 2 below. It should be remarked that the ρ_i appearing in Theorems 1 and 2 are not necessarily continuous measure functions and N is the total multiplicity of the representation. A more direct reference than [21] to the multiplicity theorem on which our results are based is to be found in the work of Plesner ([17], Theorem 10.4. 13). The version given in Stone [21] which is the one used by Hida [6] separates the representation into parts with continuous and discrete multiplicities.

We continue to assume that (1.1) holds. In what follows we write L^2 to denote the Hilbert space which is the direct sum $\sum_{i=1}^{N} \oplus L^2([0,1],\rho_i)$ and L^2_t to denote $\sum_{i=1}^{N} \oplus L^2([0,t],\rho_i)$. If $f = (f_i) \in L^2$, $f_i \in L^2([0,1],\rho_i)$ for each i and $\|f\|^2 = \sum_{i=1}^{N} \int_0^1 f_i^2(u) d\rho_i(u)$.

Theorem 2. If the Gaussian process Y is equivalent to X then its canonical representation is given by its nonanticipative representation with respect to X.

If (1.6) of Theorem A denotes the canonical

representation of X, then that of Y is given by

$$(2.20) \quad \sum_{i=1}^{N} \int_0^t [F_i(t,u) + \sum_{j=1}^{N} \int_0^t G_{ij}(u,v) F_j(t,v) d\rho_j(v)] d\xi_i(u)$$

where $[G_{ij}(u,v)]$ is a matrix-valued kernel defining a Hilbert-Schmidt operator G on L^2,

$$(2.21) \quad \sum_{i=1}^{N} \sum_{j=1}^{N} \int_0^1 \int_0^1 G^2_{ij}(u,v) d\rho_i(u) \, d\rho_j(v) < \infty \; .$$

G has the following properties:

$$(2.22) \quad G = G_1 + G_2$$

where G_1 and G_2 are Hilbert-Schmidt,

(2.23) G_1 is self-adjoint, $I + G_1$ is positive and invertible.

(2.24) G_2 is a Volterra operator,

and for some maximal chain Π containing $\{P(t)\}$,

$$(2.25) \quad G_i P = P G_i P \; , \quad (P \in \Pi)$$

for $i = 1,2$. (In (2.25) we use the same letter to denote operators which correspond to each other under the canonical isometry between $L(X)$ and L^2).
 Conversely, let G be an operator on L^2 satisfying conditions (2.21) - (2.25). Then (2.20) defines a Gaussian process equivalent to X.
Proof: The first assertion of the theorem is obvious from Theorem 1. Applying the factorization

theorem of [5] to S with respect to the chain Π of (2.25) we obtain as in (2.13), $\eta(t) = \Delta^{1/2}(I+W)X(t)$. Following the steps in [12] let us set $G_1 = \Delta^{1/2}-I$, $G_2 = \Delta^{1/2}W$ and $G = G_1 + G_2$. It follows immediately that G_1 is self-adjoint and that $(I + G_1)$ is positive and invertible. Also G_1 is Hilbert-Schmidt since $G_1 = D^{-1/2}(I - D) = D^{-1/2}(I + D^{1/2})^{-1}(I - D)$. From the properties of D or Δ we have $G_1 P = P G_1$ for every $P \in \Pi$, so that (2.25) holds for $i = 1$. Next, since Π is the common eigenchain of the Hilbert-Schmidt, Volterra operator W and of the self-adjoint operator $\Delta^{1/2}$ it follows that G_2 is Hilbert-Schmidt and has the properties (2.24) and (2.25). From (2.13)

(2.26) $\quad \eta(t) = X(t) + GX(t)$.

Since G, G_1, G_2 determine Hilbert-Schmidt operators (of the same name) on L^2, they are given by kernels and (2.21) holds where $[G_{ij}(u,v)]$ denotes the kernel of G. If ζ is any element in $L(X)$, from the canonical property (1.5) we have

(2.27) $\quad \zeta = \sum_{i=1}^{N} \int_0^1 f_i(u) \, d\xi_i(u)$,

where $f = (f_i) \in L^2$. It is easily seen that

(2.28) $\quad G\zeta = \sum_{i=1}^{N} \int_0^1 (Gf)_i(u) \, d\xi_i(u)$

$= \sum_{i=1}^{N} \int_0^1 [\sum_{j=1}^{N} \int_0^1 G_{ij}(u,v) f_j(v) d\rho_j(v)] d\xi_i(u)$.

Noting that $X(t)$ is given by (2.27) with f_i equal to the indicator function $I[o,t]$ for each i we see from (2.26) that

$$(2.29) \quad \eta(t) = \sum_{i=1}^{N} \int_0^t F_i(t,u) \, d\xi_i(u)$$

$$+ \sum_{i=1}^{N} \int_0^1 [\sum_{j=1}^{N} \int_0^t G_{ij}(u,v) F_j(t,v) d\rho_j(v)] d\xi_i(u).$$

Now for each t, $\eta(t) \in L(X;t) = \sum_{i=1}^{N} \oplus L(\xi_i;t)$.

Applying the projection operator $\sum_{i=1}^{N} P_{L(\xi_i;t)}$ to both sides of (2.29) we see that $\eta(t)$ is given by (2.20). This proves that (2.20) is the canonical representation of the process Y.

To prove the converse, suppose G, G_1 and G_2 are operators on L^2 satisfying conditions (2.21) - (2.25). Then (2.20) defines an element of $L(X;t)$ which we denote by $\eta(t)$. From condition (2.25), from the self-adjointness of G_1 and the invertibility of $I + G_1$ assumed in (2.23) it follows that $(I + G_1)^{-1}$ has Π an eigenchain. G_2 is a Volterra operator with the same property. This shows that $V = (I + G_1)^{-1} G_2$ is itself a Volterra operator with Π as an eigenchain. Setting $F = I + G_1 + G_2$ we see that $\eta(t) = F X(t)$ and $F = (I + G_1)(I + V)$ is defined on $L(X)$. Clearly, since $S = F^* F = (I + V^*)(I + G_1)^2 (I + V)$, $\eta(t)$ is the nonanticipative (and hence canonical) representation of some Gaussian process Y equivalent to X.

It should be pointed out that the canonical representation we seek is given by (2.20) where G

has to satisfy the additional conditions (2.21)-(2.25). Given an element in $L(X;t)$ of the form (2.20) these conditions are necessary and sufficient for it to be a canonical representation of an equivalent Gaussian process. The verification of (2.25) for an appropriate choice of Π seems hardest when $P(t)$ has discontinuities. However, a complete specification of G is possible in the following important special case.

Theorem 3. Let the canonical representation (1.6) of X have only continuous spectral types ρ_i. If Y is any Gaussian process equivalent to X then its canonical representation is given by

$$(2.30) \quad \sum_{i=1}^{N} \int_0^t [F_i(t,u) + \sum_{j=1}^{N} \int_u^t G_{ij}(u,v) F_j(t,v) d\rho_j(v)] d\xi_i(u)$$

or by

$$(2.30a) \quad X(t) + \sum_{i=1}^{N} \sum_{j=1}^{N} \int_0^t [\int_u^t G_{ij}(u,v) F_j(t,v) d\rho_j(v)] d\xi_i(u).$$

The G_{ij}'s satisfy (2.21) and are Volterra-type kernels.

Conversely, any element of $L(X)$ given by (2.30) (or (2.30a)) represents a Gaussian process equivalent to X.

Proof. The direct part of Theorem 2 shows that a canonical representation for Y exists and has the form (2.20). Only the matrix-kernel $[G_{ij}(u,v)]$ remains to be completely determined. From the definition of G_1 and Δ (or D) it follows that $G_1 = 0$ since the chain Π has no gaps, the ρ_i being continuous by assumption. Thus to characterize the required representation we only have to consider (2.25). Let us consider the operator $G(= G_2)$ and $P(t)$ on the Hilbert space L^2. For $f = (f_i) \in L^2$ it is easy to see that $P(t)f = I_{[0,t]}f$ ($I_{[0,t]}$ being the indicator function of $[0,t]$) and $P(t)L^2 = \{f \in L^2 : f(s) = 0$ a.e. for $t < s \leq 1\}$. Here $f(s) = 0$ a.e. for $t < s \leq 1$ means each component $f_i(s) = 0$ a.e. ρ_i for $t < s \leq 1$. Condition (2.25) says that $P(t) G P(t) = GP(t)$ for each t. It is easy to derive that for every i and j,

$$\int_0^t G_{ij}(u,v) f_j(v) d\rho_j(v) = 0 \text{ a.e. } \rho_i \text{ for } u > t$$

where $f_j \in L^2([0,1], \rho_j)$. Proceeding along the lines of the proof of Lemma 5.1 in [12] we find that $G_{ij}(u,v) = 0$ a.e. $\rho_i \times \rho_j$ for $u > v \geq 0$. The facts that the ρ_i are continuous, $\rho_i(0) = 0$ and that $\rho_1 > \ldots > \rho_N$ are used in obtaining this conclusion from which (2.30) follows immediately. The converse is also easily established if we observe that G as given in (2.30) and (2.21) where $[G_{ij}(u,v)]$ are Volterra-type kernels, is a Hilbert-Schmidt, Volterra operator which has the maximal chain $\Pi = \{P(t), 0 \leq t \leq 1\}$ for its eigenchain. Hence the converse part of Theorem 2 applies and the proof is complete.

A further special case of Theorem 3 occurs when the ξ'_i s are standard Wiener processes, i.e., $\rho_i(u) = u$. For $N = 1$, the representation (2.30) has been obtained earlier in [7], [10], and [12].

In the above theorems the emphasis has been mostly on the Hilbert space aspect of the problem and the question of the existence of stochastic processes (as families of random variables) with the requisite properties has not been examined except for the remark made towards the end of Section 1. This we now consider in some detail. As shown in Section 1, let $X(t,\omega)$ be a measurable version of the X-process defined on a complete probability space $(\Omega, \underline{A}, P)$,

$$(2.31) \quad X(t,\omega) = \sum_{i=1}^{N} \int_0^t F_i(t,u) \, d\xi_i(t,\omega) .$$

Let $\underline{B}(X;t)$ and $\underline{B}(\xi;t)$ ($0 \le t \le 1$) be the σ-fields generated respectively by the random variables $\{X(s,\omega), 0 \le s \le t\}$ and $\{\xi_i(s,\omega), 0 \le s \le t, i = 1, \ldots, N\}$ and containing all P-null sets. Clearly from (2.31), $\underline{B}(X;t) \subseteq \underline{B}(\xi;t)$ for each t and $L(X;t) \subseteq \sum_{i=1}^{N} \oplus L(\xi_i;t)$. But we must have equality in the last relation. For if $\zeta \in \sum_{i=1}^{N} \oplus L(\xi_i;t)$ and $\zeta \perp L(X;t)$ for some t we obtain

$$(2.32) \quad \sum_{i=1}^{N} \int_0^t g_i(t,u) \, F_i(s,u) \, I_{[0,s]}(u) \, d\rho_i(u) = 0$$

for all s in $[0,t]$, where $g(t,.) = (g_i(t,.)) \in L^2_t$ defines $\zeta = \sum_{i=1}^{N} \int_0^t g_i(t,u) \, d\xi_i(u)$. By the canonical property of the representation (1.6) of X the family $\{F_i(s,.), 0 \le s \le t, i = 1,\ldots,N\}$ spans L^2_t. Hence it follows that $g(t,.) = 0$ proving

$$(2.33) \quad L(X;t) = \sum_{i=1}^{N} \oplus L(\xi_i;t).$$

In fact, the argument just given is a "Hilbert space" type argument independent of which version we are dealing with, i.e., it depends on a property solely of the covariance function. As can be easily seen, from (2.33) follows that $\underline{B}(\xi;t) \subseteq \underline{B}(X;t)$ thus showing

$$(2.34) \quad \underline{B}(X;t) = \underline{B}(\xi;t)$$

for each t. If Y is the Gaussian process of Theorem 2 (or Theorem 3) with the canonical representation (2.20) (on account of the joint measurability of the kernel functions $G_{ij}(u,v)$) one can choose measurable versions of the stochastic integrals occurring in (2.20) to obtain a version of Y on $(\Omega,\underline{A},P)$ which is (t,ω)-measurable,

$$(2.35) \quad Y(t,\omega) = \sum_{i=1}^{N} \int_0^t [F_i(t,u)$$

$$+ \sum_{j=1}^{N} \int_0^t G_{ij}(u,v) F_j(t,v) d\rho_j(v)] d\xi_i(u,\omega).$$

Exactly as we did with the X-process, the canonical property of (2.20) established in Theorem 1 shows that $L(Y;t) = \sum_{i=1}^{N} \oplus L(\xi_i;t)$ and $\underline{B}(Y;t) = \underline{B}(\xi,t)$, where $\underline{B}(Y;t)$ is the σ-field generated by $\{Y(s,\omega), 0 \leq s \leq t\}$ and all P-null sets.

Theorem 4. Let X be a Gaussian process such that (1.1) is satisfied. Then

(i) A measurable version of the canonical representation exists and is given by (2.31).

(ii) Every Gaussian process Y equivalent to X has a canonical representation whose measurable version is given by (2.35). Furthermore, for each t,

(iii) $\quad L(Y;t) = \sum_{i=1}^{N} \oplus L(\xi_i;t) = L(X;t)$ and

(iv) $\quad \underline{B}(Y;t) = \underline{B}(\xi;t) = \underline{B}(X;t)$.

Theorem 4 shows that if Y is equivalent to X and (1.1) is satisfied then both processes have canonical representations expressible in terms of a sequence $\{\xi_i(t,\omega)\}$ of independent additive processes. However, it is possible for two Gaussian processes to possess the latter property without being equivalent, as is shown by the following example. Let $X(t) = B(t)$, B being a standard Wiener process and let $Y(t) = \int_0^t (t-u)dB(u)$. The right hand side representation for Y is canonical as can easily be verified. But Y and X are singular Gaussian measures because Y(t) has a version almost all of whose sample paths are differentiable.

References

[1] Cramér, H. (1964). Stochastic processes as curves in Hilbert space, <u>Teor. Veroyatnost. i. Primenen.</u>, 9, 195-204.

[2] Cramér, H. (1961). On some classes of non-stationary stochastic processes, Proc. of the 4th Berkeley Symposium in Math. Statist. and Prob., II, 57-77.

[3] Doob, J. L. (1953). Stochastic Processes, Wiley, New York.

[4] Feldman, J. (1958). Equivalence and perpendicularity of Gaussian processes, Pacific J. Math., 8, 699-708; correction, ibid. 9 (1959), 1295-1296.

[5] Gohberg, I. C. and Krein, M. G. (1970). Theory and Applications of Volterra operators in Hilbert space, (Eng. Trans.), American Math. Society, Providence, R.I.

[6] Hida, T. (1960). Canonical representations of Gaussian processes and their applications, Memoirs of the College of Science, University of Kyoto, Series A, Vol. 33, No. 1.

[7] Hitsuda, M. (1968). Representation of Gaussian processes equivalent to Wiener process, Osaka J. Math. 5, 299-312.

[8] Hitsuda, M. (1968). Multiplicity of some classes of Gaussian processes, Nagoya Math. J., 52, 39-46.

[9] Hitsuda, M. Representation of equivalent Gaussian processes. To appear.

[10] Kailath, T. and Duttweiler, D. (1972). An RKHS approach to detection and estimation problems, Part III: Generalized innovations, representations and a likelihood ratio formula. IEEE Trans. Inform. Theo. 18, 730-745.

[11] Kallianpur, G. and Oodaira, H. (1963). The equivalence and singularity of Gaussian measures. Proceedings of Symposium on Time Series Analysis, Wiley, New York, 279-291.

[12] Kallianpur, G. and Oodaira, H. (1973). Non-anticipative representations of equivalent Gaussian processes, Annals of Probability 1, 104-122.

[13] Kallianpur, G. (1973). Non-anticipative canonical representations of equivalent Gaussian processes, Multivariate Analysis -III, Academic Press, 31-44.

[14] Kallianpur, G. (1973). Canonical representations of equivalent Gaussian processes, Sankhya, Series A, 35, 405-416.

[15] Lévy, P. (1956). A special problem of Brownian motion and a general theory of Gaussian random functions., Proc. of the 3rd Berkeley Symposium on Math. Stat. and Prob., II, 133-175.

[16] Parzen, E. (1963). Probability density functionals and reproducing kernel Hilbert spaces, Proceedings of Symposium on Time Series Analysis, Wiley, New York, 155-169.

[17] Plesner, A. I. (1964). Spectral Theory of Linear Operators, Vol. II. (Eng. Trans.). Ungar, New York.

[18] Riesz, F. and Sz.-Nagy, B. (1955). Functional Analysis, Ungar, New York.

[19] Rozanov, Yu. A. (1962). The density of a Gaussian measure with respect to another one, Theory of Probability and its Appl. 7, 82-87.

[20] Siraya, T. N. (1973). On canonical representations of stochastic processes of multiplicity one and two, <u>Teor. Veroyatnost. i. Primenen.</u>, 18, 155-160.

[21] Stone, M. H. (1932). <u>Linear Transformations in Hilbert Space</u>. American Mathematical Society Colloquium Publications, V. 15.

* note added in proof; the ensuing argument can be simplified by noting that (2.12) shows that F^{-1} also has $L(X;t)$ for its invariant subspaces.

A Comparison Method for Critical Branching Processes
and an Application to Age Structure

by Peter Ney

University of Wisconsin-Madison

Summary.

A comparison method for critical branching processes introduced in [4] is briefly reviewed, and is then applied to determine the limiting age distribution of the critical age-dependent process.

1. Introduction[*] and outline of the method.

The generating function $F(s,t)$ of the standard age-dependent branching process $Z(t)$ with particle production generating function
$$f(s) = \sum_{j=0}^{\infty} p_j s^j$$
and lifetime distribution $G(t)$ is the unique bounded solution of the equation

$$(1.1) \quad F(s,t) = s[1-G(t)] + \int_0^t f[F(s,t-y)]dG(y) .$$

---

[*] I would like to acknowledge some relevant conversations with K. Athreya. For background, terminology, notation, and references see our book [1], particularly Chapter IV.

A variety of extensions to models in which particle production can occur throughout the life of the parent and/or depend on the age of the parent; to multi-type models, to immigration models, and others, lead to similar integral and functional equations. Typically the solution converges to the solution of an associated equation; in the case of (1.1)

(1.2) $$F(s,t) \to q,$$

where q is the smallest non-negative root of $s = f(s)$.

The key step in the proof of several limit laws is determination of the rate of convergence of F to q.** To this end several approaches have been traditionally used. The most obvious thing to try is to reduce (1.1) to a linear equation in $q-F(s,t) \equiv H(s,t)$, namely

(1.3) $$H(s,t) = u(s,t) + f'(q)H(s,t) * G(t),$$

**We are concerned with the asymptotics of $F(s,t)$ for each s, though uniformity in s will sometimes be important. This explicitly excludes the standard limit theorems for the super-critical case, for which one <u>a priori</u> replaces s by a suitable function of t, and then studies the behavior of the modified equation. We are here primarily concerned with critical and subcritical processes, though formally some of the results also apply to supercritical processes. There is a kind of symmetry between super and sub-critical processes, which has not, however, led to any particularly exciting results. (See e.g. Sections 8 and 11 of Chapter I of [1].)

where * is convolution in t, and where u is a complicated function depending on the unknown F. In many standard cases a priori estimates on q-F (and hence u) and careful use of renewal theory arguments yield the desired asymptotic results. When $m = f'(1) \neq 1$, we obtain either

(1.4) $\quad\quad q-F(s,t) \sim Q(s)e^{\alpha t}, \quad t \to \infty,$

where α is the root (if it exists) of

$$f'(q) \int_0^\infty e^{-\alpha t} dG(t) = 1,$$

and $Q(s) \neq 0$ under suitable hypotheses on f; or if the root α does not exist but G is in a so-called sub-exponential class, then

(1.5) $\quad\quad q-F(s,t) \sim C(s)[1 - G(t)].$

(See Chapter IV of [1] for definitions, details, and references related to these results.) A corollary of the above results in the subcritical case is the Yaglom limit law, namely the existence of

$$\lim_{t \to \infty} P\{Z(t) = k | Z(t) > 0\} = b_k \geq 0, \, k \geq 1,$$

where $\sum b_k = 1.$

In the critical case the above method can also be used to prove the exponential limit law

(1.6) $\quad\quad \lim_{t \to \infty} P\{\frac{Z(t)}{t} > z | Z(t) > 0\} = \exp\{-\frac{2\mu}{f''(1)} z\}, \, z \geq 0,$

where $\mu = \int_0^\infty t\,dG(t)$, but it requires excessive moment hypotheses. A more successful approach has been based on a comparison between $F(s,t)$ and the g.f. of the embedded generation process determined by $f(s)$. The iterates $\{F_n(s,t)\,;\,n \geq 0\}$ defined by $F_0(s,t) = s$ and

$$F_{n+1}(s,t) = s[1 - G(t)] + \int_0^t f[F_n(s,t-y)]\,dG(y), \quad n \geq 0,$$

play a central role in this analysis, and the fact that $F_n(s,t)$ is monotone in n, s, and t is used extensively in the proof. For details and references the reader is again referred to Chapter IV of [1].

In a number of natural generalizations of the standard age-dependent process one is led to equations of the form

$$(1.7) \qquad F(s,t) = \xi(s,t) + \int_0^t f[F(s,t-y)]\,dG(y)\,,$$

where $\xi(s,t)$ is a more general forcing term than that encountered in (1.1); or to a multi-dimensional version of (1.1) or (1.7). The analogs of the iterates $F_n(s,t)$ need then no longer be monotone, and the method described above breaks down (or at best becomes very complicated).

For this reason the proof of such results as the exponential limit law for the multi-type critical age-dependent process had been open for some time. Recently [4] I solved this problem by combining a very elementary new comparison idea with a well known moment argument. Since this approach avoids the careful monotonicity arguments required in earlier proofs, it turns out to be easily applicable to a great variety of critical branching models. In the remainder of this section I will review the idea behind this method, and in the next section I will illustrate it by using it to determine the limiting age distribution of a critical process.

This result can actually be derived by other simple arguments which are particular to the age problem, and which I indicate in Section 3. These do not, however, extend to other equations of the form (1.7), and my purpose here is to give an illustration of the comparision approach.

It rests on two simple observations. First that if f is a polynomial generating function (or even a g.f. of a distribution having moments of all orders), with $f'(1) = 1$, and the generating function F of the corresponding process satisfies (1.7), then all moments of this process exist. It is also easy to determine their asymptotic behavior since they satisfy ordinary (linear) renewal equations. The limiting moments of the suitably normalized and conditioned process can then be identified as those of the exponential distribution, thereby determining the limit law. This idea was first observed by H. Weiner [6], who used it to give the first proof of the exponential limit law for the standard critical age-dependent process. It has since been used by Weiner [e.g. 6,7] and others (e.g. Durham [2]) to prove these limit laws for other models. This approach, taken by itself, has the disadvantage that it requires the existence of

<u>all moments</u> of the particle production distribution.

The present idea is merely to combine the above observation with another one. Namely that the solution $F(s,t)$ of (1.7) for a process with $f'(1) = 1$ and $f''(1) < \infty$, can be approximated sufficiently closely by the solution $H(s,t)$ of an analogous equation with polynomial particle production g.f. $h(s)$. The behavior of H is then determined by the above moment method, and the conclusion is extended to F.

This kind of trick has in fact been used before. Comparison of critical branching processes with processes having linear fractional generating functions goes back a number of years (see Chapter I of [1] for references). These comparison methods however made use of particular properties of the linear fractional g.f. which do not generalize readily to more complicated models. The only novelty in the present method is the recognition that focusing on this special class of g.f.'s as a basis of comparison is misleading, and that any g.f. having moments of all orders, used in conjunction with the moment-renewal argument, will suffice.

2. <u>The age distribution of a critical age-dependent process.</u>

Let $Z(x,t)$ = the number of particles of age $\leq x$ alive at time t, in a process with particle g.f. $f(s) = \sum p_j s^j$ and lifetime distribution $G(t)$, and let $F(s,x,t)$ = the g.f. of $Z(x,t)$. It is well known (see e.g. T. Harris [3], Chapter VI) that this function is the unique bounded solution of the equation

$$(2.1) \quad F(s,x,t) = [1-G(t)][sJ(x-t) + 1 - J(x-t)] + \int_0^t f[F(s,x,t-y)]dG(y),$$

where $J(t) = 0$ for $t < 0$, $= 1$ for $t \geq 0$. Let $\mu_x = \int_0^x [1 - G(t)]t$, $\mu_\infty \equiv \mu$, and set $A(x) = \mu_x/\mu$. The reader will recognize $A(x)$ as the limiting "age distribution" in a revewal process with interoccurrence time distribution G. Note that <u>neither F, nor its approximating iterates F_n, are monotone in t.</u>

<u>Theorem.</u> <u>If $f'(1) = 1$, $f''(1) < \infty$, and $1 - G(t) = o(t^{-2})$ then</u>

$$(2.2) \quad \lim_{t \to \infty} P\{\frac{Z(t)}{t} > z | Z(t) > 0\} = \exp\{-\frac{2\mu z}{f''(1)A(x)}\}.$$

This result is of course exactly what one would expect, and when $x = \infty$ reduces to the standard result (1.6). However, the methods used in the latter case do not work here. We start with a

<u>Comparison Lemma.</u>

(i) <u>Let f be a generating function with $f'(1) = 1$ and $f''(1) < \infty$, and let $\varepsilon > 0$ be given. Then there exist polynomial generating functions $g(s)$, $h(s)$ such that $g'(1) = h'(1) = 1$ and</u>

$$f''(1) - \epsilon \leq g''(1) \leq f''(1) \leq h''(1) \leq f''(1) + \epsilon.$$

(ii) <u>If</u> f <u>and</u> h <u>are any generating functions such that</u> $f'(1) = h'(1) = 1$ <u>and</u> $f''(1) < h''(1) < \infty$, <u>then there exists an</u> s_0 <u>such that</u>

$$f(s) \leq h(s) \text{ for } s_0 \leq s \leq 1.$$

(iii) <u>Let</u> $F(s,x,t)$ <u>and</u> $H(s,x,t)$ <u>be the solutions of</u> (2.1) <u>corresponding to</u> f <u>and</u> h <u>in</u> (ii). <u>Then</u>

(2.3) $F(s,x,t) \leq H(s,x,t)$ for $s_0 \leq s \leq 1$, $0 \leq t$.

<u>Proof</u>. Part (i) is trivial, and part (ii) follows from the Taylor series

$$h(s) - f(s) = \tfrac{1}{2}(s-1)^2[h''(1) - f''(1)] + o(s-1)^2.$$

To prove (iii) let $F_0(s,x,t) = H_0(s,x,t) = s$, and define the iterates

(2.4) $F_{n+1}(s,x,t) = [1-G(t)][sJ(x-t) + 1 - J(x-t)] +$
$$\int_0^t f[F_n(s,x,t-y)]dG(y), \quad n \geq 0,$$

and similarly $\{H_n(s,x,t) ; n \geq 0\}$. Suppose that $s \geq s_0$. Then

$$F_1(s,x,t) = [1-G(t)][sJ(x-t) + 1 - J(x-t)] +$$
$$f(s)G(t)$$
$$\geq [1 - G(t)]s + f(s)G(t)$$
$$\geq s = F_0(s,x,t).$$

If $F_n \geq F_{n-1} \geq s$, then clearly $F_{n+1} \geq F_n \geq s$. Hence by induction F_n is non-decreasing in n, and $F_n(s,x,t) \geq s \geq s_0$; and similarly for H_n.

Furthermore

$$H_1(s,x,t) - F_1(s,x,t) = [h(s) - f(s)]G(t) \geq 0$$

under the hypothesis of (iii) for $s \geq s_0$, and if $H_n \geq F_n$ then substitution in (2.4) shows that $H_{n+1} \geq F_{n+1}$. Hence by induction

(2.5) $\qquad F_n(s,x,t) \leq H_n(s,x,t)$ for $n \geq 0$, $s \geq s_0$.

Since the unique bounded solution of (2.1) is the limit of the iterates in (2.4), (2.3) follows.

The next ingredient in the proof is the following

<u>Moment Lemma</u>. <u>If all moments</u> $\{f^{(k)}(1-) \; ; \; k \geq 1\}$ <u>exist, and</u> $f'(1) = 1$, <u>then</u>

(2.6) $\qquad EZ^k(x,t) \sim A(x) \dfrac{k!}{b^{k-1}} t^{k-1}, \; t \to \infty,$

where

$$b = 2\mu/f''(1)A(x)$$

<u>and</u>

$$A(x) = \int_0^x [1 - G(t)]dt / \int_0^\infty [1 - G(t)]dt.$$

The proof of this result envolves straightforward renewal theory arguments, and can be carried out entirely along the lines of the proof of the analogous result for $EZ^k(t)$. (See Weiner [6].)

Finally we will need the

<u>Extinction Probability Lemma.</u> <u>If</u> $f'(1) = 1$, $f''(1) < \infty$, <u>and</u> $\mu = \int t \, dG(t) < \infty$, <u>then</u>

(2.7) $\quad P\{Z(x,t) > 0\} \sim P\{Z(t) > 0\} \sim \dfrac{2\mu}{f''(1)t}, \quad t \to \infty.$

This says roughly that ages become sufficiently mixed, so that if there are particles of any age, then there are particles of all ages. The moment-comparison type arguments will not work to prove this kind of result, since they only give us the behavior of the g.f. $F(s,x,t)$ in the neighborhood of $s = 1$. The present lemma follows as a special case of the general theorem in [5], which I will not discuss here. An independent proof relying on the special properties of the age problem is given in Section 3 (see (3.8)).

Let us now combine the above three lemmas to prove the theorem. We start with the g.f. $f(s)$ satisfying the hypotheses of the theorem, and the associated solution $F(s,x,t)$ of (2.1). By the comparison lemma, given any $\epsilon > 0$, there is a polynomial g.f. $h(s)$ such that $h'(1) = 1$ and

$f''(1) \leq h''(1) \leq f''(1) + \epsilon$, and such that the corresponding solution H, of (2.1) satisfies $F(s,x,t) \leq H(s,x,t)$ for $s \geq s_0$. Thus for $\lambda \geq 0$

$$(2.8) \quad \frac{F(e^{-\lambda/t},x,t) - F(0,x,t)}{1 - F(0,x,t)} \leq 1 - \frac{t[1-H(0,x,t)]}{t[1-F(0,x,t)]}$$

$$+ \{\frac{H(e^{-\lambda/t},x,t)-H(0,x,t)}{1 - H(0,x,t)}\}\{\frac{t[1-H(0,x,t)]}{t[1-F(0,x,t)]}\}.$$

Let $Z_H(x,t)$ be the process associated with H, and let $b_h = 2\mu/h''(1)A(x)$. Then by the moment and extinction lemmas

$$E\{(\frac{b_h Z_H(x,t)}{t})^n | Z_H(x,t) > 0 \}$$

$$= E(\frac{b_h Z_H(x,t)}{t})^n / P\{Z_H(x,t) > 0\} \to n! \text{ as } t \to \infty.$$

Hence as in Weiner's proof in [6], we have by Carleman's theorem that

$$\lim_{t \to \infty} P\{Z_H(x,t)/t > z | Z(x,t) > 0\}$$

$$= \exp\{-b_h z\} = \exp\{- \frac{2\mu z}{h''(1)A(x)} \}.$$

By the continuity theorem for Laplace transforms this implies that

$$(2.9) \quad \lim_{t \to \infty} \frac{H(e^{-\lambda/t},x,t) - H(0,x,t)}{1 - H(0,x,t)} = \frac{1}{1 + x/b_h}.$$

Furthermore by the extinction lemma

(2.10) $$\lim_{t\to\infty} \frac{1 - H(0,x,t)}{1 - F(0,x,t)} = \frac{f''(1)}{h''(1)}.$$

Substituting (2.9) and (2.10) in (2.8), we see that

$$\overline{\lim} \frac{F(e^{-\lambda/t},x,t) - F(0,x,t)}{1 - F(0,x,t)} \le 1 - \frac{f''(1)}{f''(1) + \epsilon}$$

$$+ \{\frac{1}{1 + \frac{f''(1)A(x)}{2\mu}\lambda}\} \cdot \{\frac{f''(1)}{f''(1) + \epsilon}\}.$$

Repeating a similar argument for a lower bound, and noting that ϵ is arbitrary, we get

$$\lim \frac{F(e^{-\lambda/t},x,t) - F(0,x,t)}{1 - F(0,x,t)} = \frac{1}{1 + \frac{f''(1)A(x)}{2\mu}\lambda}.$$

This proves the theorem.

3. Other aspects of the age structure.

Throughout this section let us assume that f and G satisfy the hypothesis of the theorem. Consider $Z(x_1,t)$ and $Z(x_2,t)$ for any $0 < x_1 \le x_2 \le \infty$. The joint generating function of these processes can be shown to satisfy an integral equation similar to (2.1), and hence one can show that $\nu(x_1,x_2,t) \equiv EZ(x_1,t)Z(x_2,t)$ is the solution of

(3.1) $\nu(x_1,x_2,t) = [1-G(t)]J(x_1-t)J(x_2-t)$
$\quad + f''(1)\int_0^t \mu(x_1,t-y)\mu(x_2,t-y)dG(y)$
$\quad + \int_0^t \nu(x_1,x_2,t-y)dG(y)$.

Of course from (2.1), $\mu(x,t) = EZ(x,t)$ satisfies

(3.2) $\quad \lim_{t\to\infty} \mu(x,t) = \frac{1}{\mu}\int_0^x [1-G(t)]dt = A(x)$,

and using this in (3.1), renewal methods yield

(3.3) $\quad EZ(x_1,t)Z(x_2,t) \sim \frac{1}{\mu} f''(1)A(x_1)A(x_2)t, \quad t\to\infty$.

From this we immediately have the <u>mean square</u> convergence

(3.4) $\quad \frac{1}{t} E\left(\frac{Z(x_1,t)}{A(x_1)} - \frac{Z(x_2,t)}{A(x_2)}\right)^2 \to 0, \quad t \to \infty$.

This in turn implies the <u>convergence in probability</u>

(3.5) $\quad \frac{Z(x,t)}{Z(t)} \xrightarrow{p} A(x)$.

Namely, for any $\epsilon > 0$, $0 < c < \infty$,

$$(3.6) \quad P\{|\frac{Z(x,t)}{Z(t)} - A(x)| > \varepsilon | Z(t) > 0\}$$

$$\leq P\{|\frac{Z(x,t)}{Z(t)} - A(x)| > \varepsilon, Z(t) \geq ct\}/P\{Z(t) > 0\}$$

$$+ P\{0 < Z(t) < ct\}/P\{Z(t) > 0\}$$

$$\leq P\{|\frac{Z(x,t)}{A(x)} - Z(t)| > \frac{\varepsilon ct}{A(x)}, Z(t) \geq ct\}/P\{Z(t) > 0\}$$

$$+ P\{0 < \frac{Z(t)}{t} < c | Z(t) > 0\}$$

$$\leq P\{|\frac{Z(x,t)}{A(x)} - Z(t)| > \frac{\varepsilon ct}{A(x)}\}/P\{Z(t) > 0\}$$

$$(3.7) \quad + P\{\frac{Z(t)}{t} < c | Z(t) > 0\}.$$

Now by (1.6) the second term on the right side of (3.7) can be made arbitrarily small by taking c small. The first is bounded by

$$\frac{A^2(x)}{\varepsilon^2 c^2 t^2} E(\frac{Z(x,t)}{A(x)} - Z(t))^2 / P\{Z(t) > 0\},$$

which by (3.4) with $x_1 = x$ and $x_2 = \infty$ goes to 0 as $t \to \infty$.

The convergence of $Z(x,t)/Z(t)$ can in turn be used to prove the extinction probability lemma which we used in the last section. Just observe that for $\varepsilon < A(x)/2$

$$(3.8) \quad 1 \geq \frac{P\{Z(x,t) > 0\}}{P\{Z(t) > 0\}} \geq \frac{P\{|\frac{Z(x,t)}{Z(t)} - A(x)| < \epsilon\}}{P\{Z(t) > 0\}}$$

$$\geq P\{|\frac{Z(x,t)}{Z(t)} - A(x)| < \epsilon | Z(t) > 0\} \to 1 \text{ as } t \to \infty.$$

If higher moments (than the second) of G exist, then by using sharper forms of the rewewal theorem together with (3.1), one can sharpen (3.4) to

$$(3.9) \qquad E(\frac{Z(x_1,t)}{A(x_1)} - \frac{Z(x_2,t)}{A(x_2)})^2 = o(t^{-k})$$

where k depends on the moment of G. If G has bounded support, or even exponentially decaying tails, then I would guess that the convergence in (3.9) would be exponentially fast. More generally, one might conjecture that if $\varphi(x)$ is a function such that $\int \varphi(x) A(x) = 0$, then

$$E[\int_0^\infty \varphi(x) dZ(x,t)]^2$$

will converge to 0 at a rate like in (3.4) or (3.9) or even exponentially fast, depending on G. Note that $A(x)$ is a left eigenfunction of the mean function $M(x,y,t)$ = the expected number of particles of age $\leq y$ at t, produced by a single particle of age x at time 0 (see Harris [3], p. 153). This would be consistent with similar convergence rate

results of Athreya (see Section 7.4 of Chapter
V of [1]) for functionals of multi-type processes.

References

[1] Athreya, K. A., Ney, P.: (1972) Branching
    Processes, Springer Verlag, Berlin.
[2] Durham, S. D.: (1971) Limit theorems for a
    general critical branching process J. Appl.
    Prob. 8, 1-16.
[3] Harris, T. E.: (1963) The Theory of Branching
    Springer Verlag, Berlin
[4] Ney, P.: (1974) Critical branching process,
    To appear in J. Appl. Prob.
[5] Ney, P.: (1974) Non-linear Volterra equations
    and a Tauberian theorem, U. of Wisconsin
    technical report, Madison, Wis.
[6] Weiner, H. J.: (1965) Asymptotic properties
    of an age-dependent branching process,
    Annals of Math. Stat. 36, 1565-1568.
[7] Weiner, H. J.: (1970) An multitype critical
    age-dependent branching processes, J. Appl.
    Prob. 7, 523-543.

Stochastic Iteration of Stable Processes

by K. B. Athreya [1,2,3,4]

Abstract. Let $\{X_n(t); -\infty<t<\infty\}$ $n=1,2,\ldots$ be a sequence of i.i.d. stable processes of order α so that for each n, w.p.1. $X_n(0) = 0$, $\{X_n(t)\}$ has independent increments, and

$$|t|^{-\alpha^{-1}} X_n(t)$$

has the same distribution of $X_n(1)$. Define $Y_1(t) = X_1(t)$ and recursively $Y_{n+1}(t) = X_{n+1}(Y_n(t))$ for $n \geq 1$. It is shown here that

a) if $\alpha<1$ then $\alpha^n \log|Y_n(t)| - \log|Y_0(t)|$ converges w.p.1. to a realvalued random variable Y whose distribution is independent of t.

b) if $\alpha=1$ then $\sigma^{-1} n^{-\frac{1}{2}}(\log|Y_n(t)| - n\mu) \overset{d}{\to} N(0,1)$ where μ and σ^2 are the mean and variance of $\log|X_1(1)|$ and $\overset{d}{\to}$ means convergence in distribution.

c) if $\alpha>1$ then $\log|Y_n(t)|$ converges in distribution and the limit is independent of t.

The limiting behavior of $Y_n(t)$ and $|Y_n(t)|$ are also deduced from the above.

Keywords: Stochastic iteration, stable processes.

AMS Classification 60J30

Footnotes.

1. The work was initiated at the University of Wisconsin, Madison, U.S.A. and completed at the

K. B. ATHREYA

Institut For Matematisk Statistik, University of København, Denmark.
2. The author is on leave from the Indian Institute of Science Bangalore 560012, India.
3. This is a revised version of an invited talk given by the author at the Summer Institute on Stochastic Processes at Bloomington, Indiana, U.S.A. in July - August 1974.
4. The support from the National Science Foundation, U.S.A. and the Danish Natural Sciences Research Council, Denmark is gratefully acknowledged.

1. <u>Introduction</u>. In a recent paper [1] the concept of stochastic iteration was introduced by the author and was shown to be a generalization of the notion of branching processes. Roughly speaking the idea is this. Let $\{X_n(t,\omega); t \in T\}$ $n=1,2,\ldots$ be a sequence of stochastic processes defined on a common probability space (Ω, B, P) where the index set T is also the state space of the processes. Define $Y_1(t,\omega) = X_1(t,\omega)$ and recursively $Y_{n+1}(t,\omega) = X_{n+1}(Y_n(t,\omega),\omega)$ for $n \geq 1$. Then the sequence $\{Y_n\}$ is called the stochastic iterate of $\{X_n\}$. If the X_n's are i.i.d. random walks on the nonnegative integer lattice, then for each t the sequence $\{Y_n(t); n=1,2,\ldots\}$ is a Galton-Watson branching process with $Y_1(t)$ as the initial number and the distribution generating the random walk as the offspring distribution. If the X_n's are i.i.d. random walks on the whole integer lattice then for each t the sequence $\{Y_n(t)\}$ is a selfannihilating branching process (see [3]). If the X_n's are i.i.d. processes on $[0,\infty)$ with stationary and independent nonnegative increments then for

each t the sequence $\{Y_n(t)\}$ is a continuous state space branching process with $Y_1(t)$ as the initial amount. One can go on like this and realize branching processes in random environments, branching processes with state dependent population growth and so on as stochastic iterates. It is well known (see [1]) that in most branching processes context the population either dies out or explodes and there is no stability. A natural question is what happens to the stochastic iterate $\{Y_n\}$ of a sequence $\{X_n\}$ more general than the ones mentioned in the above examples. Specifically, what is the limiting behavior of $\{Y_n\}$. We answer this question fairly completely when the X_n's are i.i.d. stable processes on $(-\infty, \infty)$. There is a trichotomy depending on the order α of the process. There is stability if and only if $\alpha > 1$. For $\alpha < 1$,

$|Y_n|^{\alpha^n}$ has a limit and for $\alpha=1$ $(|Y_n|e^{-n\mu})^{n^{-\frac{1}{2}}}$

is asymptotically normal for some appropriate μ. For $\alpha > 1$, $Y_n(t)$ has a limit distribution independent of $t \neq 0$.

2. <u>Statement of the results</u>. Let $\{X_n(t,\omega); -\infty < t < \infty\}$ be a sequence of i.i.d. processes defined by the conditions

i) $\{X_1(t,\omega); t \geq 0\}$ has stationary independent increments, $X_1(0,\omega) = 0$ w.p.1. and for any $t>0$, $t^{-\alpha^{-1}} X_1(t)$ has a distribution independent of t.

ii) $\{X_1(-t,\omega); t \geq 0\}$ is an independent copy of $\{X_1(t); t \geq 0\}$.

These hypotheses imply that the characteristic function of $X_1(t)$ is stable and is of the form $\exp(t\psi(\theta))$ where

$$\psi(\theta) = \begin{cases} ia\theta - c|\theta|^\alpha \{1+i\beta\frac{\theta}{|\theta|}\tan\frac{\pi\alpha}{2}\} & \text{if } \alpha \neq 1 \\ ia\theta - c|\theta|^\alpha \{1+i\beta\frac{\theta}{|\theta|}\frac{2}{\pi}\log|\theta|\} & \text{if } \alpha = 1 \end{cases}$$

with a and β real, $|\beta| \leq 1$, $c > 0$, $0 < \alpha \leq 2$. Further since $t^{-\alpha^{-1}} X_1(t)$ has the same distribution as $X_1(1)$ the function $\psi(\theta)$ must satisfy $t\psi(\theta t^{-\alpha^{-1}}) = \psi(\theta)$ for all θ real and $t > 0$.

This imposes the following constraints on a and β

i) If $\alpha \neq 1$ then $a = 0$ and hence if $\alpha = 2$ then $\{\frac{X_1(t,\omega)}{c} ; t \geq 0\}$ is a standard Brownian motion process. Note, however, that if $\alpha \neq 1$ and $\alpha < 2$ then the process need not even be symmetric about the origin.

ii) If $\alpha = 1$ then $\beta = 0$ but a need not vanish. Thus $\{X_1(t) ; t \geq 0\}$ is a noncentered Cauchy process with drift at and a scale coeffecient c. Thus $\{\frac{X_1(t) - at}{c} ; t \geq 0\}$ is a standard Cauchy process.

If $\alpha \geq 1$ then the support of the probability distribution of $X_1(1)$ is the whole real line. This is so since the density function is analytic at least on the strip $|\text{Im}Z| < c$. If $\alpha \leq 1$ and $|\beta| = 1$ then the distribution of $X_1(1)$ is one-sided (bounded on the left if $\beta = -1$ and on the right if $\beta = +1$). For proofs of those observation see [4]. Notice that these are consistent with our results. If the distribution of

$X_1(1)$ was one-sided then from the classical branching process theory we would expect instability for $\{Y_n\}$. This is indeed the case even if $X_1(1)$ were not one-sided so long as $\alpha<1$. We are now ready to state our results.

Theorem 1. Let $0<\alpha<1$. Then,

i) $\lim_{n\to\infty} |Y_n(t,\omega)|$ exists w.p.1., but assumes only two values namely 0 and ∞.

ii) $\lim_{n\to\infty} |Y_n(t,\omega)|^{\alpha^n}$ exists w.p.1. and equals $\exp\{|t|+Z(t,\omega)\}$ where $Z(t,\omega)$ is a real valued random variable having an absolutely continuous distribution that is identical to that of

$\sum_1^\infty \alpha^j \mu_j$ where $\{\mu_j: j=1,2,\ldots\}$ are i.i.d. as $\log|X_1(1)|$. Thus, the distribution of $Z(t,\omega)$ is independent of t.

iii) On the set $\{\omega: |t|+Z(t,\omega)>0\}$ $\lim_n |Y_n(t,\omega)| = \infty$ but $\{Y_n(t,\omega)\}$ need not converge if $P\{X_1(1)<0\}P\{X_1(1)>0\} > 0$. On the set $\{\omega: |t|+Z(t,\omega)<0\}$ $\lim_n |Y_n(t,\omega)| = 0$ and hence $Y_n(t,\omega) \to 0$ also.

Theorem 2. Let $\alpha=1$ and $\mu=E\log|X_1(1)|$. Then;

i) $\lim_{n\to\infty} |Y_n(t,\omega)| = \begin{cases} \infty & \text{w.p.1. if } \mu>0 \\ 0 & \text{w.p.1. if } \mu<0 \end{cases}$

ii) If $\mu=0$ then w.p.1. both 0 and ∞ are limit points for the sequence $\{|Y_n(t,\omega)|\}$.

iii) There exist no normalising sequence c_n such that $c_n |Y_n|$ converges w.p.1. or in law to a nondegenerate proper limit distribution.

iv) $(|Y_n|e^{-n\mu})^{n^{-2^{-1}}} \xrightarrow{d} e^{N\sigma}$ where N is a standard normal random variable, σ^2 the variance of $\log|X_1(1)|$ and $\xrightarrow{d}$ means convergence in distribution.

<u>Theorem 3</u>. Let $\alpha > 1$. Then, $Y_n(t,\omega) \xrightarrow{d} Y$ where the random variable Y has a distribution function independent of t given by

$$P(Y \leq y) = \int_0^\infty G(y|x|^{-\alpha}) \, dH(x)$$

where $G(x) = P(X_1(1) \leq x)$, $H(x) = P\{\exp(\sum_{j=1}^\infty \alpha^{-j} \mu_j) \leq x\}$

the μ_j's being i.i.d. as $\log|X_1(1)|$. Further, the sequence $\{Y_n(t,\omega)\}$ cannot converge with probability one.

3. <u>Proof of the theorems</u>. Fix $t \neq 0$. Then since the X_n's are stable processes we conclude that $Y_n(t,\omega) \neq 0$ for any n. Thus, we may take logarithm of $|Y_n|$ to yield

$$Z_{n+1}(t,\omega) = \eta_{n+1}(t,\omega) + \alpha^{-1} Z_n(t,\omega) \qquad (1)$$

where $Z_n(t,\omega) = \log|Y_n(t,\omega)|$,

$$\eta_{n+1}(t,\omega) = \log \left| \frac{X_{n+1}(Y_n(t,\omega),\omega)}{|Y_n(t,\omega)|^{\alpha^{-1}}} \right|$$

for n=0,1,2,... (By definition, $Y_0(t,\omega) = t$ w.p.1.) Suppressing t and ω for convenience and iterating

(1) yields

$$Z_{n+1} = \sum_{j=0}^{n} \alpha^{-j} \eta_{n+1-j} + \alpha^{-(n+1)} \log|t|. \qquad (2)$$

Let $F_n \equiv \sigma(Y_j(t); j=1,2,\ldots,n)$ be the sub-σ-algebra generated by $Y_j(t)$ $j=1,2,\ldots,n$ of the basic σ-algebra $\acute{B}$ of the triplet $(\Omega, \acute{B}, P)$ on which all the stochastic processes mentioned so far are defined. It is clear that the random variables $\{\eta_j\}$ are adapted to this family $\{F_j\}$ and that the conditional distribution of η_{j+1} given F_j is independent of the conditioning and the same as that of $\log|X_1(1)|$. This yields the following

Lemma 1. Fix $t \neq 0$. The random variables $\{\eta_j(t,\omega); j=1,2,\ldots\}$ are i.i.d. as $\log|X_1(1)|$ where the η_j's are as in (1).

With this background we now proceed to the proofs of the three theorems.

Proof of Theorem 1. Here $\alpha < 1$. Multiply both sides of (2) by α^{n+1} to get

$$\alpha^{n+1} Z_{n+1} = \sum_{j=1}^{n} \alpha^j \eta_j + \log|t|. \qquad (3)$$

Since the η_j's are identically distributed and since $\log|X_1(1)|$ has a finite mean (in fact, all its moments are finite) we get

$$E\left(\sum_{j=1}^{\infty} \alpha^j |\eta_j|\right) = (E|\eta_1|)(1-\alpha)^{-1} < \infty \text{ and hence that}$$

$\sum_{j=1}^{\infty} \alpha^j \eta_j$ converges w.p.1. This proves part (ii) of

Theorem 1. Part (i) is a trivial consequence of part (ii). Part (iii) follows by noting that under the hypothesis $P(X_1(1)>0)P(X_1(1)<0) > 0$ the sets $\{\omega: Y_n(t,\omega) \text{ is ultimately positive}\}$ and $\{\omega: Y_n(t,\omega) \text{ is ultimately negative}\}$ have probability zero.

Proof of Theorem 2. Here $\alpha=1$. Equation (2) yields

$$Z_{n+1} = \sum_{j=1}^{n+1} \eta_j + \log|t|.$$

The random variable $\log|X_1(1)|$ has all moments and in particular the mean and the variance. Part (i) easily follows by the strong law of large numbers. When $\mu=0$, the law of the iterated logarithm asserts that w.p.1. the sequence

$$\left\{ \frac{1}{\sqrt{2n\log\log n}} \sum_{j=1}^{n} \eta_j : n=1,2,\ldots \right\}$$

has the entire interval $[-\sigma,+\sigma]$ as the set of its limit points. This shows (ii) and the fact that the sequence $\{Z_n\}$ does not converge at all. To prove (iii) we just note that since the η_j's are i.i.d. with finite mean and variance there cannot exist constant constants d_n such that $\sum_{j=1}^{n} \eta_j - d_n$ converges in law or w.p.1.

Finally part (iv) follows from the central limit theorem.

Proof of Theorem 3. Here $\alpha>1$. Equation (2) and the fact that the X_n's are i.i.d. yields the conclusion that Z_{n+1} has the same distribution as

$$Z^*_{n+1} = \sum_{j=1}^{n} \alpha^{-j}\eta_j + \alpha^{-(n+1)}\log|t|.$$

As in the proof of Theorem 1, the series $\sum_{j=1}^{\infty} \alpha^{-j}\eta_j$ converges w.p.1. and the sequence Z^*_{n+1} converges w.p.1. to $\sum_{j=1}^{\infty} \alpha^{-j}\eta_j$. This shows that $|Y_n|$ converges in distribution to $H(\cdot)$. Now note that

$$P(Y_{n+1} \leq y) = P\left(\frac{X_{n+1}(Y_n)}{|Y_n|^{\frac{1}{\alpha}}} \leq y |Y_n|^{-\frac{1}{\alpha}}\right)$$

$$= \int_0^{\infty} G(y|x|^{-\frac{1}{\alpha}}) dP(|Y_n| \leq x).$$

The function $G(y|x|^{-\frac{1}{\alpha}})$ is continuous for x in $(0,\infty)$ and bounded. Also $|Y_n| \xrightarrow{d} H$. Thus $Y_n \xrightarrow{d} F$ where F is defined in Theorem 3. The limiting distribution F is easily seen to be nondegenerate. Now $\{Y_n\}$ being a Markov Chain having a nondegenerate limiting distribution cannot converge with probability one.

REFERENCES

1. K. B. Athreya (1974) "Stochastic Iteration" MRC Technical Summary Report #973, University of Wisconsin, Madison, Wisconsin U.S.A. 53706.
2. K. B. Athreya and P. Ney (1972) Branching Processes, Springer - Verlag, Heidelberg.
3. K. B. Erickson (1973) "Selfannihilating branching processes" Ann. Prob. Vol. 1, No. 6, 926-946.

4. E. Lukacs (1960), <u>Characteristic Functions</u>, No. 5, Griffin's Statistical Monographs and Course, Charles Griffin & Co., London.

THE ERGODIC BEHAVIOR OF A
CLASS OF REAL TRANSFORMATIONS
by J. H. B. Kemperman
University of Rochester

0. Summary. Section 1 is concerned with applicable sufficient conditions for a transformation $T: R \to R$ to be exact. In Section 2 we consider the ergodic behavior of the sequence of iterates $\{T^n x\}$ when $x \in R$ and T belongs to a certain class of meromorphic functions. Proofs and further results may be found in [1] and [2]. There is also a discussion of some open problems.

1. Exact endomorphisms. Let T be a given transformation of the measure space $R = (R, \mathcal{B}, \lambda)$. Here, λ denotes Lebesgue measure and $\mathcal{B}$ the σ-field of all Lebesgue measurable subsets of R. The domain D of T is assumed to be an open set such that $\lambda(D^c) = 0$. Let

(1) $$D = \bigcup_{s \in S} I_s$$

be the unique decomposition of D into disjoint non-empty open intervals. We further assume that, for all $s \in S$, the restriction $T|I_s$ has a continuous non-zero derivative and a full range $R_s = R$. Let $f_s: R \to I_s$ denote the inverse of this restriction. Finally assume that $2 \leq |S| \leq \infty$ with $|S|$ as the number of elements in the index set S.

EXAMPLE. Let each interval I_s be finite with midpoint m_s and width $c_s \pi$, thus, $I_s = \{x \in R: |x - m_s| < c_s \pi/2\}$ where $c_s > 0$. Define

(2) $$Tx = a_s \tan(\frac{x - m_s}{c_s}) + b_s \text{ for all } x \in I_s.$$

Here the a_s and b_s denote real constants depending on $s \in S$, $a_s \neq 0$.

Inspired by ideas due to Rényi [4] and Rohlin [5], we derived [1] some very general conditions which are sufficient for T to have the following property.

(*) The transformation T admits at most one invariant probability measure ν which is equivalent to λ. Moreover, if such a measure ν exists then the endomorphism T of the measure space (R, ß, ν) is <u>exact</u> and thus mixing of all orders and thus ergodic.

REMARK. The partition (1) is a generator for T as soon as there exists a T-invariant probability measure ν on ß equivalent to λ. More precisely, for each point $x \in R$ such that $T^n x$ is well-defined for all $n \geq 0$ (as is true for almost all x) the value x is uniquely determined by its sequence of "digits" $\{s_k(x), k \geq 0\}$; here, $s = s_k(x)$ is the value $s \in S$ such that $T^k x \in I_s$, (k = 0, 1, ...).

The "exactness" of the endomorphism T is the analogue of the Kolmogorov property for an invertible transformation T. Namely, T is exact when the tail σ-field determined by $\{T^n x\}$ is trivial. Equivalently, the endomorphism T is exact when its natural extension to an automorphism (see [5], p. 23) is a Kolmogorov automorphism. One way of representing this natural extension would be as a 1:1 shift in the doubly infinite product $\ldots \times S_{-1} \times S_0 \times S_1 \times \ldots$ with S_j as a copy of S. Exactness may also be defined by the requirement that $\lim_n \nu(T^n A) = 1$ for each $A \in \mathcal{B}$ with $\nu(A) > 0$, see [5] p. 11.

Further observe that the required finite invariant measure ν would have a density $\psi = d\nu/d\lambda$ satisfying

(3) $\qquad \psi(x) > 0; \int_{-\infty}^{+\infty} \psi(x) \, dx < \infty,$

and further the functional equation

(4) $\qquad \sum_{Tx = y} \psi(x) \left| (\frac{d}{dx}) Tx \right|^{-1} = \psi(y),$

(for almost all $y \in R$).

DEFINITION. Let ϕ be a measurable function on R such that

(5) $\qquad \phi(x) > 0; \int_{-\infty}^{+\infty} \phi(x) \, dx < \infty.$

For C as a subinterval of R, let $K_\phi(C) \leq +\infty$ denote the smallest positive constant such that

(6) $\phi(x) \phi(T^n x)^{-1} \left| (\frac{d}{dx}) T^n x \right|^{-1}$

$\leq K_\phi(C) \phi(y) \phi(T^n y)^{-1} \left| (\frac{d}{dy}) T^n y \right|^{-1}$

holds for each choice of the positive integer n and each choice of the points x and y in the domain of T^n such that $T^n x \in C$, $T^n y \in C$ and further, for

$j = 0, 1, \ldots, n-1$, $s_j(x) = s_j(y)$, (that is, $T^j x$ and $T^j y$ belong to the same interval I_{s_j}).

THEOREM 1. *Suppose there exists a measurable function ϕ satisfying (5) such that $K_\phi(C) < \infty$ for each compact subinterval C of R. Then T satisfies* (*). *Moreover, $K_\phi(R) < \infty$ is sufficient for the existence of a T-invariant probability measure ν equivalent to λ.*

As an application of Theorem 1 we have the following result. Here, K_N is defined by

$$(7) \qquad K_N = \inf \left| \left(\frac{d}{dx}\right) T^N x \right|,$$

where x runs through the domain of T^N, ($N = 1, 2, \ldots$).

THEOREM 2. *The following conditions imply that* (**) *the transformation T admits a unique invariant probability measure ν equivalent to λ; moreover, T is exact relative to ν.*

  (i) *We have $\sup_s \|I_s\| < \infty$ with $\|I_s\|$ as the length of the interval I_s. Moreover, $K_1 > 0$ while $K_N > 1$ for some $N \geq 1$.*

  (ii) *The functions $f_s: R \to I_s (s \in S)$ are of class C^2 and satisfy*

$$(8) \qquad \sup_s \sup_x |f_s''(x)/f_s'(x)| < \infty.$$

  (iii) *There exists a measurable function $\phi: R \to R$ of class C^1 satisfying (5) and such that*

$$(9) \qquad \sup_x |\phi'(x)/\phi(x)| < \infty$$

and moreover,

(10) $\sup_s \sup_x \sup_y |f'_s(x)/f'_s(y)|\{\phi(y)/\phi(x)\} < \infty$.

COROLLARY. Property (**) holds for the transformation T defined by (2) provided all the sequences $\{a_s\}$, $\{1/a_s\}$, $\{b_s\}$ and $\{c_s\}$ are bounded and, moreover, $K_N > 1$ for some $N \geq 1$.

Proof. Apply Theorem 2 with $\phi(x) = 1/(1 + x^2)$.

Let us now turn to the special case of (2) where T is defined by

(11) $Tx = a(\tan x) + b$

when x is not an odd multiple of $\pi/2$. Here, a and b are real constants, $a \neq 0$. Thus, taking S as the set of all integers, we have $I_s = \{x \in R: |x - s\pi| < \pi/2\}$, $m_s = s\pi$, $c_s = 1$, $a_s = a$ and $b_s = b$, for all $s \in S$. One may as well assume that $|b| \leq \pi/2$ and further that T has no stable fixed point $z_0 \in R$, that is, a real value z_0 such that $a(\tan z_0) + b = z_0$ and $|a|\sec^2 z_0 < 1$. Equivalently we assume that either $|a| > 1$, $|b| \leq \pi/2$ or $0 < |a| \leq 1$ and

(12) $b_0 \leq |b| \leq \pi/2$; here,

$b_0 = \gamma - (\text{sgn } a)(\sin \gamma)(\cos \gamma)$.

Further γ is defined by $(\cos \gamma)^2 = |a|$ and $0 \leq \gamma < \pi/2$.

Since $K_1 = |a|$, we see that (**) holds as soon as $|a| > 1$. One can further show $K_2 > 1$ (and hence (**)) holds when $|a| \leq 1$ and $|b| > \gamma + (\sin \gamma)(\cos \gamma)$. This includes the case $-1 \leq a < 0$ and $b_0 < |b| \leq \pi/2$.

2. A class of meromorphic transformations.

In the present Section, we restrict ourselves to the special case where $T: D \to R$ with $\lambda(D^c) = 0$ is of the special form $Tx = g(x)$ $(x \in D)$ with $g(z)$ as a meromorphic function defined for all complex z and such that $\text{Im}\{g(z)\}$ has a constant sign $\epsilon = \pm 1$ in the upper half plane $\text{Im}\{z\} > 0$. Equivalently, the meromorphic function g has only real and simple poles c_s and is of the form

$$(13) \quad g(z) = A + \epsilon[Bz - \sum_s p_s(\frac{1}{z - c_s} + \frac{1}{c_s})].$$

Here, A, B and p_s are real constants such that $B \geq 0$, $p_s > 0$ and $\sum_s p_s/c_s^2 < \infty$; further $\{c_s\}$ has no finite accumulation points.

Let us also assume that $|B| < 1$ and further that g does not have any <u>real</u> fixed point z_0 with $|g'(z_0)| \leq 1$.

It turns out [2] that then the transformation T on R has a unique invariant probability measure ν which is equivalent to λ. Moreover, T is mixing relative to ν and thus ergodic. Next, the density $\psi = d\nu/d\lambda$ is a Cauchy density of the form

$$(14) \quad \psi(x) = (q/\pi)\{(x - p)^2 + q^2\}^{-1},$$

with p and q as real constants, $q > 0$. Finally, $z_0 = p + iq$ is precisely the unique <u>complex</u> fixed point in the upper half plane of either the transformation $z' = g(z)$ or $z' = \overline{g(z)}$, depending on whether $\epsilon = +1$ or $\epsilon = -1$, respectively.

The proof that the above function $\psi(x)$ does

satisfy the functional equation (4) makes use of the integral

$$(2\pi i)^{-1} \int_L \psi(z) \, (g(z) - y)^{-1} dz$$

(with L as a line parallel to the real axis) and some residue calculus. In fact, substantially the same proof shows that the T-transform of a Cauchy distribution of type $z = p + iq$ is precisely the Cauchy distribution of type $g(p + iq)$ or $\overline{g(p + iq)}$, depending on whether $\varepsilon = +1$ or $\varepsilon = -1$.

The following more transparent proof of this curious property was kindly pointed out to me by Professor Harry Kesten. Consider a planar Brownian motion $\{X(t)\}$ with starting point $X(0) = p + iq$. It hits the real axis for the first time at a point $X(T)$ having precisely the Cauchy distribution (14). Observe further that the many-to-one transformation $g(z)$ is everywhere analytic apart from some poles on the real axis and is such that $g(z)$ is real if and only if z is real. We infer that the image process $\{X'(t) = g(X(t))\}$ is a Brownian motion apart from a time substitution (P. Lévy) with $X'(0) = g(p + iq)$ as the starting point and further that $X'(T) = g(x(T))$ is precisely the place where the image process hits the real axis for the first time. Consequently, $g(X(T))$ must have a Cauchy distribution with parameters p' and q' such that $p' + i\varepsilon q' = g(p+iq)$.

In the special case (11) the above results imply that T is mixing relative to a unique invariant Cauchy distribution ν as soon as $|a| > 1$

and also when $0 < |a| \leq 1$ and $b_o < |b| \leq \pi/2$, compare (12). It seems likely that in all these cases T is not only mixing but even exact relative to ν. In view of the results at the end of Section 1, it remains only to consider the case that

$$0 < a \leq 1; \quad \gamma - (\sin \gamma)(\cos \gamma) < |b| \leq \gamma + (\sin \gamma)(\cos \gamma).$$

Similar conjectures can be formulated for the general function (13). Since no "classical" 1:1 transformation is known which is Kolmogorov and not Bernoulli (see [3] p. 110) one might guess that the natural extension of the endomorphism T ([5] p. 22) is not only mixing and not only Kolmogorov but even isomorphic to a Bernoulli shift (with the same entropy).

Also under investigation are the situations where T possesses a real fixed point z_o with $|g'(z_o)| = 1$, (as happens for instance in the special case where $Tx = \tan x$). Then T admits an infinite invariant measure ν equivalent to λ with density $(x - z_o)^{-2}$. It is likely that: (i) T is ergodic relative to ν; (ii) for almost all starting points x most of the iterates $T^n x$ are very close to x_o; (iii) let U and V be closed sets with $x_o \notin U \subset V$; then for almost all starting points x the event $T^n x \in U$ has among the events $T^n x \in V$ a relative frequency equal to $\nu(U)/\nu(V)$. In fact, (i) would imply both (ii) and (iii).

There exists a large literature on the asymtotic behavior of a sequence of iterates $\{g_n(z)\}$ determined by a given rational, meromorphic or entire function $g(z)$ of a complex variable z.

Let $\mathfrak{J}(g)$ denote the invariant set consisting of all complex numbers z' such that $\{g_n(.)\}$ is not a normal family in any neighborhood of z'. Usually one studies for the points $z \notin \mathfrak{J}(g)$ the convergence of $\{g_n(z)\}$ to the different stable fixed points or fixed cycles.

However, in studying invariant measures one would rather be interested in starting points $z \in \mathfrak{J}(g)$. Particularly intriguing seem those entire functions g for which $\mathfrak{J}(g)$ is the entire complex plane. Already Fatou conjectured that this might be true for the function $g(z) = e^z$. Thus it would not seem inconceivable that the very classical transformation $z' = e^z$ admits an invariant measure equivalent to planar Lebesgue measure.

Quite different from (11) seems to be the transformation on R defined by $Tx = a(\sec x)$, with $a \neq 0$ as a real constant. Here, it can easily happen that T has no real fixed point x with $|g'(x)| \leq 1$ while nevertheless T^3 does have a stable real fixed point; (for instance, take $a = 1.009805$ ... as the root of the equation $a[\sec\{a(\sec a)\}] = -\pi$). Possibly there are always powers T^N of T with a stable fixed point $x \in R$ so that T would not have any invariant measure equivalent to Lebesgue measure.

REFERENCES

[1] J. H. B. Kemperman, Applicable sufficient conditions for the existence of a finite invariant measure for a piecewise monotone transformation, to appear.

[2] J. H. B. Kemperman, The ergodic behavior of a class of meromorphic transformations related to the tangent function, to appear.

[3] D. S. Ornstein, Ergodic theory, randomness, and dynamical systems Yale Mathematical Monographs no. 5, Yale University Press 1974.

[4] A. Rényi, Representations for real numbers and their ergodic properties, Acta Math. Hungar. 8 (1957) 477-493.

[5] V. A. Rohlin, Exact endomorphisms of a Lebesgue space. Amer. Math. Soc. Translations (2) 39 (1964) 1-36.

Mixing Conditions and Asymptotic Theory of Stationary Time Series[*]

by I. Zhurbenko
Statistical Laboratory
Moscow State University

The asymptotic theory of independent random variables is well developed [1]. The purpose of the present paper is to discuss the possibility of extending all the main results which are known for independent random variables to the case of stationary processes satisfying certain mixing conditions. The conditions given by Rosenblatt in [2] in 1957 are by now very well known. They assume

$$\alpha(\tau) = \sup_{t, B_1 \in \beta(-\infty, t]} |P(B_1 \times B_2) - P(B_1)P(B_2)| \to 0, \quad B_2 \in \beta[t + \tau, +\infty),$$

when $\tau \to \infty$, where $\beta[\tau, \infty)$ is a σ-algebra generated by x_t, $t \in [\tau, \infty)$. It was shown, however, in [3] that these mixing conditions are not sufficient to obtain some of the asymptotic results for time series.

---

[*]This paper was written while the author was visiting the Department of Statistics and the Statistical Laboratory, University of California, Berkeley. This paper was partially supported by NIH Research Grant GM10525, National Institute of Health, Public Health Service.

Here we make use of the new mixing conditions of Rjauba [4] and Statulevicius [5] in which

$$\beta(\tau) = \sup \left| P\left(\frac{A}{B_1 B_2}\right) - P\left(\frac{A}{B_1}\right) \right| \to 0, \quad t, A \in \beta[t],$$
$$B_1 \in \beta(t, t+\tau),$$
$$B_2 \in \beta[t+\tau],$$

if $\tau \to \infty$. These may be called mixing conditions of Markov type. The following result was proved in [3].

Theorem 1. Consider a stationary time series x_t with discrete time t such that $|x_t| < C_1$ with probability one and such that

$$\beta(\tau) \le C_2 \exp\{-\beta\tau\}.$$

Then the moment-generating function of $S_n = \sum_{i=1}^{n} x_t$ satisfies the equality

$$\ln E \exp\{zS_n\} = n\psi(Z) + \psi_1(Z) + R_n(Z), \qquad (1)$$

where $\psi(Z)$, $\psi_1(Z)$, and $R_n(Z)$ are analytic functions of complex parameter Z in some neighborhood $|Z| < r$. The function $R_n(Z)$ is such that

$$|R_n(Z)| \le \exp\{-\frac{\beta n}{2}\}$$

uniformly in $|Z| < r$. The constants r, K depend only on C_1, C_2, β.

It is interesting to note that in the case of independent identically distributed random variables

$$\psi(it) = \ln E \exp\{itX_1\},$$
$$\psi_1(Z) \equiv 0, \qquad R_n(Z) \equiv 0.$$

We can obtain an interesting example concerning the moving average of independent identically distributed random variables x_t. Let

$$y_t = \sum_k x_{t+k}\, \alpha_k,$$

where α_k is a given sequence of real numbers such that $\alpha_k = \alpha_{-k}$ and $\sum_k |\alpha_k| < \infty$. The random variable y_t exists with probability one. We have

$$S_{2n+1} = \sum_{t=-n}^{n} y_t = \sum_{t=-n}^{n} \sum_k x_{t+k}\, \alpha_k$$
$$= \sum_{t=-n}^{n} x_t \sum_k \alpha_k + \sum_{k=1}^{\infty} \sum_{p=0}^{\infty} \alpha_{k+p} \{x_{n+k} + x_{-n-k} - x_{n+1-k}$$
$$- x_{-n-1+k}\}.$$

It is not difficult to find the functions $\psi(Z)$ and $\psi_1(Z)$ of the asymptotic formulae (1) for this example. In fact $\exp\{\psi(it)\}$ is the characteristic function of $W = x_1 \sum_k \alpha_k$ and $\exp\{\frac{1}{4}\psi_1(it)\}$ is the characteristic function of random variable

$$W_1 = \sum_{k=1}^{\infty} \sum_{p=0}^{\infty} \alpha_{k+p} \{x_{n+k}\}.$$

We suppose x_1 has a symmetric distribution of mean zero. We have to note that the random variable W_1 exists with probability one if the series

$$\sum_{n=1}^{\infty} n|\alpha_n|$$

converges. The random variables W and W_1 are independent, but we cannot say the same about the various terms of the right-hand side of the last equation in (2). We can say only that expansion (2) is more suggestive than formula (1). We note that in the case of a stationary Gaussian process y_t, the expansion $y_t = \sum_k \alpha_t x_{t+k}$ exists under simple restrictions on the correlation function of y_t. The random variables W, W_1 are independent Gaussian random variables.

Generally, we can define the function $\psi(z)$ and $\psi_1(Z)$ in the following way:

$$\psi(z) = \lim_{n \to \infty} \frac{\ln E \exp\{zS_n\}}{n},$$

$$\psi_1(z) = \lim_{n \to \infty}(\ln E \exp\{zS_n\} - n\psi(z)),$$

whenever they exist. Theorem 1 gives us an example of a class of stationary time series for which these two limits exist in the neighborhood of the point $z = 0$. Let us suppose that these limits exist for any point of the line $z = it$, t a real parameter. It is easy to show then that $|\exp\{\psi(it)\}| \leq 1$ for any real t, $\psi(0) = \psi_1(0) = 0$. In our example with the moving average the functions $\psi(it)$ and $\psi_1(it)$ are the logarithms of the characteristic functions of the random variables which we constructed above. In the general case of stationary time series the functions $\psi(z)$ and $\psi_1(z)$ are not necessarily logarithms of characteristic functions. It is possible that the Fourier transform of the function $\exp\{\psi(it)\}$ may be negative at some points. Nevertheless, we can investigate the asymptotic behavior of S_n in terms of the functions $\psi(it)$ and $\psi_1(it)$ in much the same way as for independent identically distributed random variables. We can find a simple example in which $\psi(it)$ is not a characteristic function. The equality $\psi''(0) = 0$ means that the variance of our signed measure is equal to zero. A necessary condition is that $\sum_t \text{Cov}(x_0 x_t) = 0$. But, at the same time, it is very easy to give an example for which $\psi^{(4)}(0) \neq 0$, and we know that this is impossible for distributions with nonnegative probability measure. The limit problems, in which one has to consider convolutions of more general functions of bounded variation than the distribution functions were already discussed in [6].

There is the possibility $\psi(it) \equiv 0$. In our example with the moving average this well be the case when $\sum_k \alpha_k = 0$. The asymptotic behavior of S_n depends on $\psi_1(it)$; that is, the distribution of S_n converges to the distribution function with characteristic function $\exp\{\psi_1(it)\}$ and the rate of convergence is determined by R_n. In the paper [7], for the more general case of stationary processes, it was shown that either $\sup \text{Var}(S_n) < \infty$ or $\text{Var } S_n = nb(n)$, where $b(n)$ is a slowly varying (Karamata) function. In our case, $\lim_{n\to\infty} b(n) = \psi''(0)$ and $\lim_{n\to\infty} \text{Var } S_n = \psi_1''(0)$ if $\psi''(0) = 0$.

We can obtain another interesting example if we put $\alpha_k = \frac{1}{2^{|k|}}$ for the moving average with independent identically distributed random variables such that $P\{x_t = 1\} = \frac{1}{2}$ and $P\{x_t = -1\} = \frac{1}{2}$. It is easy to show that in this case y_t is uniformly distributed on some interval. In this case, $\psi_1(it)$ is the logarithm of characteristic function of the random variable $x_t \sum_k \frac{1}{2^{|k|}}$ and it is periodic in view of the distribution of x_t. Then, in spite of the fact that y_n is continuously distributed, the local limit theorems for

$$\sum_{i=1}^{n} y_n$$

are described in a discrete form.

A number of limit theorems for stationary time series with mixing conditions have been proved in [1], [2], [4], [5], [7]. The present paper gives us the possibility of proving different limit theorems for stationary processes using only the ordinary techniques of independent random variables applied to the function $\psi(it)$.

References

1. I. Ibragimov and Yu. Linnik, <u>Stationary and independent sequences of random variables</u> (1970).
2. M. Rosenblatt, A central limit theorem and a strong mixing condition. <u>Proc. Nat. Acad. Sci. U.S.</u> <u>42</u> (1956), 43-47.
3. I. Zhurbenko, Strong estimates for mixed semi-invariants of random processes. <u>Siberian Math. J.</u> <u>13</u> (1972), 202-213.
4. B. Rjauba, The central limit theorem for sums of series of weakly dependent random variables. <u>Litovsk. Math. Sb.</u> <u>2</u> (1962).
5. V. Statulevicius, Limit theorems for dependent random variables under various regularity

conditions. A lecture delivered at the Int. Cong. Math., Vancouver, Canada (August 1974).
6. H. Bergstrom, _Limit theorems for convolutions_, Stockholm, Almqvist & Wiksel; New York, Wiley (1963).
7. I Ibragimov, Remark on the central limit theorems for dependent random variables. _Teor. Verojatn. i Primenen_. (To appear.)

Martingales and Diffusion Processes
Satisfying Boundary Conditions

by Walter A. Rosenkrantz
University of Massachusetts

I. Introduction.

Martingales of the form $U(t,x(t))$ where $x(t)$ is a one dimensional time homogeneous diffusion process and $U(t,x)$ is a suitable function have been studied by several authors including Doob [5], Arbib [1], Stroock-Varadhan [10] and Lai [6]. The simplest and possibly best known example of such a martingale occurs when $x(t)$ is the Wiener process and $U(t,x) = x^2-t$. Our main result is that for a large class of one dimensional diffusion processes $x(t)$ and suitable functions $v(t,x)$ the stochastic process

(1) $\quad z(t) = v(t,x(t)) - \int_0^t [v_s(s,x(s)) + Gv(s,x(s))]ds$

is a martingale. Here G is the strong infinitesimal generator of the contraction semigroup $T(t)f(x) = E_x f(x(t))$. If we set $v(t,x) = x^2-t$ and define $Gf(x) = f''(x)/2$, so G is the infinitesimal generator of the Wiener process $w(t)$, then it is easily established that $v_s(s,x) + Gv(s,x) = 0$. Thus "proving" that $v(t,w(t)) = w^2(t)-t$ is a martingale. In fact Lai op. cit. has given a general theorem to the effect that if $U_t(t,x) + GU(t,x) = 0$ then $U(t,x(t))$

is a **Martingale**. This is not always true, of course, as the following simple counter example shows. Let $x(t) = |w(t)|$ and set $U(t,x) = x$. Then $x(t)$ is the Wiener process with a reflecting barrier at the origin, $Gf(x) = f''(x)/2$, $U_t(t,x) + GU(t,x) = 0$ but $U(t,x(t)) = |w(t)|$ is clearly not a martingale. The reason for this is that $f(x) = x$ is <u>not</u> in the domain of the infinitesimal generator G which requires that $f'(0) = 0$. This last example illustrates the necessity of restricting our attention to those functions $v(t,x) \in D(G)$, the domain of G, for every $t \geq 0$. In addition our approach, using only the most basic facts of semigroup theory, allows us to construct martingales associated with diffusion processes $x(t)$ satisfying boundary conditions e.g. the Weiner process with a reflecting boundary at the origin. Other examples are discussed in part III. Doob and Stroock-Varadhan assume that $Gf(x) = a(x)f''(x) + b(x)f'(x)$ where $a(x) > 0$ and $a(x)$ and $b(x)$ are bounded and continuous. These assumptions rule out all "generalized second order differential operators of the Feller type" $G = D_m D_p^+$ (for the definition and properties of such operators the reader is referred to Mandl [7], Chapter II). The methods of this paper on the contrary are applicable to these more general operators.

Acknowledgement: In deriving the results presented here the author benefited from a useful conversation with Professor D. Stroock of the University of Colorado.

II. Construction of Martingales.

It is convenient to assume that the state space of $x(t)$ is an interval $I = (c,d)$ where $-\infty \le c < 0 < d \le +\infty$. In addition we assume that $x(t)$ has continuous paths and is strongly Markov in which case its strong infinitesimal generator $Gf = s\text{-}\lim_{t \to 0^+} (T(t)f-f)/t$ is of the Feller type $G = D_m D_p^+$. The functions $u(x) = \int_0^x m(s)dp(s)$, $v(x) = \int_0^x p(s)dm(s)$ play an important role in the study of the boundary behavior of the process. Here we assume that $\lim_{x \to c} u(x) = \lim_{x \to d} u(x) = +\infty$ i.e. the boundaries are "inaccessible". In part III we discuss examples where one or both of the boundaries are accessible. We are now ready to construct martingales of type(1).

Remark: The proof has been adapted from Stroock [9], p. 34.

Theorem 1: Suppose $v(t,x)$ is continuously differentiable with respect to the t-variable i.e. $v_t(t,x)$ is jointly continuous in t and x, and that as a function of $x \in I$, $v(t,x) \in D(G)$ and $v_t(t,x) \in D(G)$ for each $t \ge 0$. Then

$$z(t) = v(t,x(t)) - \int_0^t [v_s(s,x(s)) + Gv(s,x(s))]ds$$

is a martingale.

Remark: If $f \in D(G)$ then $v(t,x) = T(t)f(x)$ satisfies the conditions of theorem 1.

Proof: Set $F(t) = B(x(u), 0 \le u \le t)$ i.e. $F(t)$ is the smallest sigma field with respect to which the random variables $x(u)$, $0 \le u \le t$ are measurable.

It is most convenient to sketch the proof first --putting in the necessary details afterwards.

Step 1: The following decompostion holds for $t_1 < t_2$:

$$E\{v(t_2, x(t_2)) - v(t_1, x(t_1)) | F(t_1)\} = I_1 + I_2 + I_3$$

where

$$I_1 = E\{\int_{t_1}^{t_2} [v_s(s, x(s)) + Gv(s, x(s))] ds | F(t_1)\}$$

$$I_2 = E\{\int_{t_1}^{t_2} [v_s(s, x(t_2)) - v_s(s, x(s))] ds | F(t_1)\}$$

$$I_3 = E\{\int_{t_1}^{t_2} [Gv(t_1, x(s)) - Gv(s, x(s))] ds | F(t_1)\}.$$

This decompostion will be derived below. Assuming, for the moment, its validity we come to

Step 2: $I_2 + I_3 = 0$.

Deferring the proof of this as well we conclude that

(3) $E\{v(t_2, x(t_2)) - v(t_1, x(t_1)) | F(t_1)\} = I_1$

where I_1 can be rewritten as

(4) $I_1 = E\{\int_0^{t_2} [v_s(s, x(s)) + Gv(s, x(s))] ds$

$- \int_0^{t_1} [v_s(s, x(s)) + Gv(s, x(s))] ds | F(t_1)\}.$

But this is clearly equivalent to the assertion that $E\{Z(t_2) | F(t_1)\} = Z(t_1)$, which is the martingale property. We now proceed to prove the assertions contained in steps 1 and 2. We first recall the well known fact that for $f \in D(G)$ $f(x(t)) - \int_0^t Gf(x(s))ds$ is a martingale. In particular then so is the process $v(t_1, x(t_2)) - \int_0^{t_2} Gv(t_1, x(s))ds$. Thus

(5) $E\{v(t_1, x(t_2)) - \int_0^{t_2} Gv(t_1, x(s))ds | F(t_1)\}$

$= v(t_1, x(t_1)) - \int_0^{t_1} Gv(t_1, x(s))ds.$

Next observe - taking into account the obvious cancellations - that

$$I_1 + I_2 + I_3 = E\{\int_{t_1}^{t_2} v_s(s, x(t_2))ds | F(t_1)\}$$
$$+ E\{\int_{t_1}^{t_2} Gv(t_1, x(s))ds | F(t_1)\}.$$

But,

(6) $E\{\int_{t_1}^{t_2} v_s(s, x(t_2))ds | F(t_1)\} =$

$E\{v(t_2, x(t_2)) - v(t_1, x(t_2)) | F(t_1)\},$

and from (5),

(7) $E\{\int_{t_1}^{t_2} Gv(t_1, x(s))ds | F(t_1)\} =$

$E\{v(t_1, x(t_2)) - v(t_1, x(t_1)) | F(t_1)\}.$

Adding (6) and (7) yields the decomposition of step 1. We now complete the proof of theorem 1 by showing that $I_2 + I_3 = 0$.

We first show that

(8) $\quad I_2 = E\{\int_{t_1}^{t_2} \int_s^{t_2} Gv_s(s,x(u))du\,ds\,|F(t_1)\}$

$\quad\quad = \int_{t_1}^{t_2} E\{\int_s^{t_2} Gv_s(s,x(u))du\,ds\,|F(t_1)\}$

$\quad\quad = \int_{t_1}^{t_2} E\{v_s(s,x(t_2))-v_s(s,x(s))\,|F(t_1)\}ds$

$\quad\quad = E\{\int_{t_1}^{t_2}[v_s(s,x(t_2))-v_s(s,x(s))]ds\,|F(t_2)\}$

On the other hand interchanging the order of integration in (8) yields

$\quad I_2 = E\{\int_{t_1}^{t_2} \int_{t_1}^{u} Gv_s(s,x(u))ds\,du\,|F(t_1)\}$

$\quad\quad = E\{\int_{t_1}^{t_2} \int_{t_1}^{u} \frac{\partial}{\partial s} Gv(s,x(u))ds\,du\,|F(t_1)\}$

$\quad\quad = E\{\int_{t_1}^{t_2}[Gv(u,x(u))-Gv(t_1,x(u))]du\,|F(t_1)\}$

$\quad\quad = -I_3. \quad\quad\quad\text{Q.E.D.}$

Corollary: If in addition to the conditions of theorem 1 $v(t,x)$ satisfies the equation

(9) $\quad v_t(t,x) + Gv(t,x) = 0$

then $v(t,x(t))$ is a martingale.

Theorem 1 as it now stands does not by itself include all the interesting applications, but is easily extended to cover the concrete examples discussed by Lai as well as some others which he has not. To simplify matters we assume both boundaries c and d are natural. Then the domain $D(G)$ consists of all functions f with the property that f and Gf are in $C(I)$, the space of bounded continuous functions on I - the <u>boundary points included</u>.

Let $g(x,\lambda)$ and $h(x,\lambda)$ denote the monotone increasing and decreasing solutions of the equation $Gf(x) = \lambda f(x)$, $\lambda > 0$. In particular then $\lim_{x \to d} g(x,\lambda) = +\infty$, $\lim_{x \to c} h(x,\lambda) = +\infty$. Let $v(t,x)$ denote either $\exp(-\lambda t)g(x,\lambda)$ or $\exp(-\lambda t)h(x,\lambda)$. Then proceeding in a purely formal manner we have $Gv(t,x) = \lambda v(t,x)$ and $v_t(t,x) = -\lambda v(t,x)$ so $v_t(t,x)+Gv(t,x) = 0$. Hence, by our corollary, $v(t,x(t))$ is a martingale (this is part of Lai's Theorem 5, p. 428). There is however one flaw in this argument and it is that $v(t,x)$ as a function of x is unbounded and therefore is not in $D(G)$, so Theorem 1 cannot be applied. This difficulty can be circumvented by means of the following lemma, the proof of which is in the appendix to this paper.

Lemma 1: Suppose f and $D_m D_p^+ f$ are continuous on the open interval (c,d) but are not necessarily bounded. Then for any compact subinterval $[a,b] \subset (c,d)$ there exists a function Φ with the following properties:

(1) Φ and $D_m D_p^+ \Phi$ are in $C(I)$,

(2) $\Phi(x) \equiv f(x)$ for $a \le x \le b$.

In particular if c and d are inaccessible boundaries then $\Phi \in D(D_m D_p^+)$. Of course $D_m D_p^+ \Phi = f$ on $[a,b]$. We continue now with our proof of Theorem 1. Pick an expanding sequence of intervals $[a_n, b_n] \subset [a_{n+1}, b_{n+1}]$ whose union is (c,d). Let $g(x,\lambda) = f(x)$ and $[a_n, b_n] = [a,b]$ in lemma 1 and denote the corresponding function Φ by $\Phi_n(x)$. Clearly $\Phi_n \in D(D_m D_p^+)$ and so the function $U(t,x) = \exp(-\lambda t)\Phi_n(x)$ (n will be held fixed for the time being) satisfies the condition of theorem 1. In particular $U_t(t,x) + GU(t,x) = 0$ for $a_n \le x \le b_n$. Let τ_n denote the first passage time of the process $x(t)$ to either of the boundary points a_n or b_n. Then we have

(10) $U_t(t, x(t \wedge \tau_n)) + GU(t, x(t \wedge \tau_n)) = 0$

If we now apply Doob's optional stopping theorem to the martingale

$$U(t, x(t)) - \int_0^t [U_s(s, x(s)) + G U(s, x(s))] ds$$

we conclude that

$$U(t \wedge \tau_n, x(t \wedge \tau_n)) = \exp(-\lambda(t \wedge \tau_n)) g(x(t \wedge \tau_n))$$

is a martingale. In particular then we have

(11) $g(x, \lambda) = E_x(\exp(-\lambda(t \wedge \tau_n)) g(x(t \wedge \tau_n), \lambda))$.

We would like to let n increase to infinity and pass to the limit in (11) and obtain

(12) $g(x,\lambda) = E_x(\exp(-\lambda t)g(x(t),\lambda))$.

From this equation one concludes immediately that $\exp(-\lambda t)g(x(t),\lambda)$ is a martingale. We shall only sketch the proof here, for from this point on the argument is the same as Lai's on p. 426 lemma 6. First we write

$E_x(\exp(-\lambda(t \wedge \tau_n)))\ g(x(t \wedge \tau_n),\lambda)) =$

$E_x(\exp(-\lambda t)g(x(t),\lambda);\ \tau_n \geq t)\ +$

$E_x(\exp(-\lambda \tau_n)g(x(\tau_n),\lambda);\ \tau_n \leq t)$.

The proof is completed by then showing that the second summand converges to zero as $n \to \infty$ and that the first summand converges to $E_x(\exp(-\lambda t)g(x(t),\lambda))$. It is perhaps worth while reminding the reader that our goal here is not merely to reproduce Lai's results but to apply our methods to diffusion processes satisfying boundary conditions i.e. to the case where either c or d is accessible. The reason why our methods work in this case is to be found in the hypotheses of theorem 1, where we assume $v(t,x) \in D(G)$. If the reader has followed the discussion up to this point then he need hardly be reminded that $D(G)$ is determined by the boundary conditions that are imposed. In general $D(G)$ is difficult to characterize but for operators of the Feller type

$D_m D_p^+$ all possible domains $D(D_m D_p^+)$ are known - see Mandl [7]. We shall pursue this line of thought further and in more detail in part III.

III. Application of the preceeding results to diffusion processes satisfying boundary conditions.

As our first example we shall consider the so-called "Bessel processes of index $\gamma+1$". By this is meant a diffusion process $x(t)$ whose state space is the positive half line $R_+ = [0,\infty)$ and whose infinitesimal generator G has the form

(13) $\quad Gf(x) = (1/2)f''(x) + (\gamma/x)f'(x), \quad -1/2 < \gamma.$

When $\gamma = (n-1)/2$, n an integer, then $x(t)$ is the radial component of n dimensional Brownian motion. In all cases ∞ is a natural boundary but 0 is a regular or entrance boundary according as $-1/2 < \gamma < 1/2$ or $1/2 \le \gamma$. The case $\gamma \ge 1/2$ is essentially contained in Theorem 6 in p. 434 of Lai, op. cit. So we shall confine ourselves to the case - not treated by Lai - where 0 is a regular reflecting boundary i.e., $-1/2 < \gamma < 1/2$. The reflecting boundary condition is characterized analytically by defining $D(G)$ in a suitable way. This has been done in [4], [8] and we restate those results here. Let $C_0(R^+)$ denote the Banach space of bounded continuous functions f, with domain $R^+ = [0,\infty)$ and such that $\lim_{x \to \infty} f(x) = 0$. A routine application of the Hille-Yosida theorem (see either [4], [7] or [8] for the details) yields the result that the differential operator G defined at (13) is the infinitesimal generator of a strongly continuous contraction semi-group T(t) with domain D(G) defined by the conditions

(14) $D(G) = \{f : f \in C_0(R^+), Gf \in C_0(R^+), f'(0) = 0\}$

The Markov process $x(t)$ associated with this semi-group, i.e. $T(t)f(x) = E_x f(x(t))$, will be called the Bessel process of index $\gamma+1$ with a reflecting boundary at 0. N.B. we are assuming $-1/2 < \gamma < 1/2$. Let $g(x,\lambda)$ denote the increasing solution to the equation

(15) $(1/2)g''(x,\lambda) + (\gamma/x)g'(x;\lambda) = \lambda g(x,\lambda), \quad \lambda > 0.$

An explicit solution in terms of the "modified Bessel functions" can be given. A useful reference here is the book of Birkhoff-Rota [2] pp. 226-228.

The modified Bessel function $I_\nu(y)$ of order ν is defined by the power series

(16) $I_\nu(y) = (y/2)^\nu \sum_{k=0}^{\infty} \frac{(y/2)^{2k}}{k!\,\Gamma(\nu+k+1)}.$

Moreover I_ν is a solution to the differential equation

(17) $x^2 I_\nu''(x) + x I_\nu'(x) - (x^2 + \nu^2)I_\nu(x) = 0.$

The reader may now easily verify for himself that the function $f(x) = x^{-\nu} I_\nu(cx)$ satisfies the equation

(18) $f''(x) + [(2\nu+1)/x]f'(x) = c^2 f(x).$

Thus $g(x,\lambda) = x^{1/2-\gamma} I_{\gamma-1/2}((2\lambda)^{1/2}x)$ is the increasing solution to the differential equation (15) and satisfies the boundary condition $g'(0,\lambda) = 0$. Lai has shown p. 430 op. cit. that for $1/2 \leq \gamma$ the stochastic process $\exp(-\lambda t)g(x(t),\lambda)$ is a martingale. Now when $-1/2 < \gamma < 1/2$, 0 is an accessible boundary and neither of Lai's theorems 5 and 6 cover this case. Nevertheless, we have

Theorem 2: $\exp(-\lambda t)g(x(t),\lambda)$ is a martingale for all $\gamma > -1/2$.

Proof: Let $\Phi_n(x) \in C_0^\infty(R^+)$ with the property that $\Phi_n(x) \equiv 1$ for $0 \leq x \leq n$. Then

$$g_n(x,\lambda) = g(x,\lambda)\Phi_n(x) \in D(G)$$

and if

$$U(t,x) = \exp(-\lambda t)g_n(x,\lambda)$$

then it is easily verified that $U_t(t,x) + GU(t,x) = 0$, $0 \leq x \leq n$. Now let τ_n denote the first passage time to the point n. Proceeding in a manner by now familiar we conclude that

$$\exp(-\lambda(t\wedge\tau_n))g(x(t\wedge\tau_n),\lambda)$$

is a martingale. As we saw in part II in order to prove that $\exp(-\lambda t)g(x(t),\lambda)$ is a martingale we must justify the passage to the limit. In particular it suffices to show that $\lim_{n\to\infty} g(n,\lambda)E_x(\exp(-\lambda\tau_n)) = 0$.

But this can be done exactly as in Lai, op. cit. p. 426. Q.E.D.

As another example let us do exercise 10 on p. 365 of Breiman's book [3] via the martingale methods of this paper. Let [a,b] be a compact subinterval of the state space (c,d) and let τ denote the first passage time of the diffusion process x(t) with infinitesimal generator $D_m D_p^+$ to the boundary of [a,b]. Then the following identity holds

$$(19) \quad E_x\{\int_0^\tau f(x(s))ds\} = \int_a^b G(x,y)f(y)dm(y),$$

where f is a bounded continuous function on [a,b] and

$$G(x,y) = v_-(x)v_+(y)/w, \quad x \leq y$$
$$= v_-(y)v_+(x)/w, \quad y \leq x$$

where $v_-(x) = p(x) - p(a)$; $v_+(x) = p(b) - p(x)$ and $w = (p(b) - p(a))^{-1}$. Set

$$F(x) = \int_a^b G(x,y)f(y)dm(y).$$

It is an instructive exercise for the reader to verify that

$$D_m D_p^+ F(x) = -f(x) \text{ and } F(a) = F(b) = 0.$$

Let x(t) denote the diffusion process absorbed when it hits either of the boundaries a or b. Then F is in the domain of $D_m D_p^+$ and hence $F(x(t)) + \int_0^t f(x(s))ds$ is a martingale. In particular

$$E_x(F(x(\tau)) + \int_0^\tau f(x(s))ds) = F(x) = \int_a^b G(x,y)f(y)$$

But $x(\tau) = a$ or b and therefore $F(x(\tau)) = 0$. This completes the proof of (19).

Appendix.
Proof of Lemma 1.

Pick $\delta > 0$ and then choose continuous functions $f_0(x)$, $f_1(x)$ with the property that

(A.1) $\int_{a-\delta}^{a} f_0(x)dm(x) = +1 = \int_b^{b+\delta} f_1(x)dm(x)$.

In addition we can assume that f_0 and f_1 vanish on the complements of the intervals $[a-\delta, a]$, $[b, b+\delta]$ respectively. Set $j(x) = \int_c^x [f_0(y) - f_1(y)]dm(y)$. The function has the useful properties listed below:

(A.2)
i) $j(x) \equiv 1$, $a \le x \le b$
ii) $j(x) = 0$, $x \notin [a-\delta, b+\delta]$
iii) $0 \le j(x) \le 1$, all x.

Set $h(x) = \int_c^x j(y)dp(y)$ and note that $h \in D(D_m D_p^+)$. Observe that for $a \le x \le b$ we can write $h(x) = p(x) - p(a) + c'$, $c' = \int_{a-\delta}^a j(y)dp(y)$ and that $h(x) = 0$ for $c \le x \le a-\delta$. Thus

(A.3) $p(x) = h(x) + c''$, $c'' = p(a) - c'$ on $[a,b]$.

We assume as usual that $c < 0 < d$, and $p(0) = 0 = m(0^+) = m(0)$. Then

$$f(x) = f(0) + D_p^+ f(0)p(x) + \int_0^x \int_0^y D_m D_p^+ f(s)dm(s)dp(y)$$

Since $D_m D_p^+ f(s)$ is continuous on (c,d) it is possible to choose a sequence of continuous functions $g_n(x)$ vanishing in a neighborhood of the boundary points c, d with the property that $\lim_{n\to\infty} g_n(x) = D_m D_p^+ f(x)$ uniformly on a compact interval $[a',b'] \supset [a,b]$. It is clear that we can even choose the sequence g_n in such a way so that $\lim_{n\to\infty} g_n(x) = g(x)$ uniformly on $[c,d]$. Set

$$\Phi_n(x) = f(0) + D_p^+ f(0) p(x) + \int_0^x \int_0^y g_n(s) dm(s) dp(y).$$

Clearly $\lim_{n\to\infty} \Phi_n(x) = \Phi(x)$ exists and $\Phi(x) = f(0) + D_p^+ f(0) p(x) + \int_0^x \int_0^y g(s) dm(s) dp(y)$. Now by (A.3) we have $p(x) = h(x) + c''$ for $a \le x \le b$. If we now define

$$\Phi(x) = f(0) + D_p^+ f(0)(h(x)+c) + \int_0^x \int_0^y g(s) dm(s) dp(y)$$

then this function satisfies the requirements of the lemma since $D_m D_p^+ \Phi(x) = g(x) = f(x)$ on $[a,b]$. This completes the proof.

References

[1] M. Arbib, Hitting and Martingale characterizations of one dimensional diffusions, Zeit. Fur Wahr. und verw. Gebeite 4(1965), 232-247.

[2] G. Birkhoff and G. C. Rota, Ordinary Differential Equations, Ginn and Company (1960).

[3] L. Breiman, Probability, Addison-Wesley Reading, Massachusetts (1968).

[4] H. Brezis, W. Rosenkrantz and B. Singer, On a degenerate elliptic parabolic equation occurring in the theory of probability, Comm. on

Pure and Applied Math. Vol. XXIV, 395-416 (1971).

[5] J. L. Doob, Martingales and one dimensional diffusion, Trans. Amer. Math. Soc. 78, 168-208, (1955).

[6] T. L. Lai, Space-time Processes, Parabolic functions and one dimensional diffusions, Trans. Amer. Math. Soc. 175, 409-438 (1973).

[7] P. Mandl, Analytical treatment of one-dimensional Markov processes, Springer-Verlag, New York (1968).

[8] W. Rosenkrantz, An application of the Hille-Yosida theorem to the construction of martingales, Indiana Univ. Math. Journal, Nov. 1974 - to appear.

[9] D. Stroock, Diffusion Processes with continuous coefficients, Lecture notes, Univ. of Minnesota, Institute of Technology, School of Mathematics, Minneapolis, Minnesota, 55455.

[10] D. Stroock and S. Varadhan, Diffusion Processes with continuous coefficients, I, Comm. on Pure and Applied Math. Vol. XXI, 345-400 (1969).

Random Fields Generated by Quantum Fields[*]

by Victor Goodman

Indiana University

Introduction. Generalized random fields which satisfy a certain multidimensional Markov property have been introduced by Nelson [6] in order to construct operator families forming models of quantum fields. The Markov condition is an analog of one given by Pitt [9] for random fields. In this paper we derive necessary conditions for a Gaussian generalized random field of a certain type to have the Markov property. The work was suggested by Pitt's work for Gaussian process, and although our result is perhaps not the most general, the condition for the Markov property is readily investigated. We apply the result to compare Nelson's correspondence between Markov and quantum fields with a more general correspondence formulated in [8].

Definition. Let D denote an open subset of R^n. $S(R^n)$ denotes the space of rapidly decreasing C^∞ real-valued functions on R^n, [3] definition 1.7.1 , and we set

$$S(D) = \{f \in S(R^n): \text{suppf} \subset D\}$$

where suppf denotes the support of f, $\overline{\{x: f(x) \neq 0\}}$.

---

[*] Research supported by NSF Grant GP-38514.

$S(D)$ has the relativized topology of $S(R^n)$.

Definition. Let $D \subset R^n$ be open. A <u>generalized random field</u> ϕ over D is a linear map from $S(D)$ into a set of random variables over some probability space which is continuous in probability. That is, if $f_n \to f$ in the topology of $S(D)$, then $\phi(f_n) \to \phi(f)$ in measure. Such a field is said to be <u>Gaussian</u> if each $\phi(f)$ is a Gaussian random variable.

Remark: If ϕ is Gaussian, then ϕ is continuous from $S(D)$ into the space of square-integrable random variables. Hence, the covariance form $R(\cdot,\cdot)$ defined by

$$R(f,g) = E[(\phi(f)-E\phi(f))(\phi(g)-E\phi(g))]$$

is continuous in each argument.

Definition. Let ϕ be a generalized random field over D. If $U \subset D$ is open, $\mathfrak{F}(U)$ denotes the Borel subfield of the associated probability space generated by the random variables $\phi(f)$ where $f \in S(U)$.

If $A \subset D$ is closed, the Borel field $\mathfrak{F}(A)$ is defined by

$$\mathfrak{F}(A) = \bigcap \{\mathfrak{F}(U): U \text{ open}, A \subset U\}.$$

Definition. (Nelson, Pitt) A generalized random field ϕ over D satisfies the <u>Markov property</u> iff, for each open subset U of D and each integrable random variable u measurable with respect to $\mathfrak{F}(U')$, $U' \equiv D-U$,

$$E\{u|\mathfrak{F}(\overline{U})\} = E\{u|\mathfrak{F}(\partial U)\}.$$

Remark: In [7] and [9] it is pointed out that the above definition is weaker than the usual notion of

Markov, even in the case of $D \subset R^1$.

Throughout the remainder of this paper we will be concerned only with Gaussian generalized random fields with mean zero. That is, $E[\phi(f)]=0$, $f \in S(D)$.

Gaussian Generalized Random Fields.

Definition. A covariance form $R(.,.)$ over $S(D)$ is **strictly proper** iff. there exists a continuous linear map $T: S(D) \to S(D)$ with dense range such that

$$R(Tf,g) = \langle f,g \rangle_2 \quad \text{for all} \quad f,g \in S(D).$$

Here, $\langle .,. \rangle_2$ denotes the L^2 inner product relative to Lebesgue measure on D.

Remark: a mean zero Gaussian field ϕ with a strickly proper covariance is proper in the sense of Gel'fand and Vilenkin [2], for if $f \in S(D)$ and $f \neq 0$, then $\langle f,f \rangle_2 \neq 0 \Rightarrow R(Tf,f) \neq 0 \Rightarrow R(f,f) > 0 \Rightarrow \phi(f) \neq 0$. Also, the map T is unique. Suppose a linear map S satisfies the same identity as T. Then

$$0 = R(Tf,g) - R(Sf,g)$$
$$= R(Tf-Sf,g).$$

We set $g=Tf-Sf$ and conclude from a comment above that $g=0$.

Definition. Let $R(.,.)$ be strictly proper. Since R is a strickly positive inner product on $S(D)$, the norm $\|f\|_R \equiv R(f,f)^{\frac{1}{2}}$ determines a metric. Let $H(D)$ denote the real Hillbert space obtained by taking the metric space completion of $S(D)$.

If $U \subset D$ is open, we set $H(U) = \overline{S(U)}^{\|\cdot\|_R}$ and if $A \subset D$ is closed,

$$H(A) \equiv \bigcap \{H(U): \text{ U open, } A \subset U\}.$$

Definition: A map $T: \mathcal{S}(D) \to \mathcal{S}(D)$ is local iff, for each open subset U of D, supp $(Tf) \subset \bar{U}$ for each $f \in \mathcal{S}(D)$ such that supp$f \subset \bar{U}$.

Theorem. The mean zero Gaussian generalized field over $D \subset R^n$ corresponding to a strickly proper covariance R satisfying $R(Tf,g) = <f,g>_2$ has the Markov property only if

(i) for each open subset U of D,
$$H(\bar{U}) \cap H(U') = H(\partial U).$$

(ii) T is local.

The proof follows from a sequence of lemmas.

Lemma 1. Let ϕ be a strickly proper field. For a closed subset A of D, let P denote the orthogonal projection of H(D) onto H(A). Then

$$E\{\phi(F) \mid \mathfrak{F}(A)\} = \phi(PF) \qquad \text{for} \quad F \in H(D).$$

proof. If U is any open neighborhood of A, $PF \in H(U)$ and hence is an $\|\cdot\|_{R_2}$ limit of functions in $\mathcal{S}(U)$. Hence, $\phi(PF)$ is an L^2 limit of random variables measurable relative to $\mathfrak{F}(U)$ and thus is measurable relative to $\mathfrak{F}(U)$. Therefore, $\phi(PF)$ is measurable relative to $\mathfrak{F}(A)$.

Suppose $G \in H(A)$. Then $E[\phi(G)\phi((I-P)F)] = R(G, (I-P)F) = 0$. Given a collection $\{G_j\}_1^m \subset H(A)$, consideration of the above covariance, where G ranges over linear combinations of the collection yields that $\phi((I-P)F)$ is independent of $\{\phi(G_j)\}_1^m$. Hence, if $H: R^m \to R$ is a bounded Borel measurable function,
$$E[\phi((I-P)F)H(\phi(G_1), \cdots, \phi(G_m))] = 0.$$

Since $\phi(F) = \phi(PF) + \phi((I-P)F)$, it follows immediately from the defining property of conditional expectation that

$$\phi(PF) = E\{\phi(F) \mid \mathfrak{F}(A)\}.$$

Corollary: If $A \subset D$ is closed and $F \in H(D)$, then $\phi(F)$ is $\mathfrak{F}(A)$ measurable iff. $F \in H(A)$.

Lemma 2. If a strickly proper Gaussian field over D has the Markov property, then for any open subset U of D, the projections of $H(D)$ onto the spaces $H(U')$, $H(\overline{U})$ commute.

proof. Suppose U is open and $F \in H(D)$. Let Π, P denote the projections onto $H(U')$, $H(\overline{U})$. Since $\Pi F \in H(U')$, from the corollary above we have $\phi(\Pi F)$ is $\mathfrak{F}(U')$ measurable. Then from the Markov property we have

$$E\{\phi(\Pi F) \mid \mathfrak{F}(\overline{U})\} = E\{\phi(\Pi F) \mid \mathfrak{F}(\partial U)\}.$$

The left hand expression is $\phi(P\Pi F)$ from lemma 1. However, since the right hand expression is measurable relative to $\mathfrak{F}(\partial U) \subset \mathfrak{F}(U')$, the corollary above implies

$$\Pi P \Pi F = P \Pi F.$$

Hence, $\Pi P \Pi = P \Pi$. We take adjoints to get $\Pi P \Pi = \Pi P$ and thus $\Pi P = P \Pi$.

Corollary: If a strickly proper Gaussian field has the Markov property then for any open subset U of D,

$$H(\overline{U}) \cap H(U') = H(\partial U).$$

Proof. From the definition of $H(\partial U)$, $H(\partial U) \subset H(\overline{U}) \cap H(U')$. Let G be an element of the latter intersection, and let Π, P denote the obvious projections. Then $\Pi G = G$, $PG = G$ and thus, $P\Pi G = G$. From the proof of the lemma, for any $F \in H(D)$, $\phi(P\Pi F)$ is measurable relative to $\mathcal{F}(\partial U)$. From the corollary to lemma 1, $P\Pi F \in H(\partial U)$. Hence $G \in H(\partial U)$.

Lemma 3. Let $R(.,.)$ denote a covariance determined by a strickly proper Gaussian field and let T satisfy $R(T f, g) = <f, g>_2$. If U is open and supp $f \subset \overline{U}$, $f \in \mathcal{S}(D)$, then

$$Tf \text{ is orthogonal to } H(W')$$

where W is any open neighborhood of $\overline{U}$.

proof. Suppose $g \in \mathcal{S}(D)$ and supp $g \subset U'$. Since $f = g = 0$ on ∂U, we have

$$R(Tf, g) = <f, g>_2 = 0.$$

Given an open neighborhood W of $\overline{U}$, choose an open set V such that $\overline{U} \subset V \subset \overline{V} \subset W$. Then $\overline{V}'$ is an open neighborhood of W'. If $g \in \mathcal{S}(D)$ and supp $g \subset \overline{V}'$ then supp $g \subset U'$ and hence Tf is orthogonal to g. But, such elements g are dense in $H(W')$ and therefore Tf is orthogonal to $H(W')$.

proof of the Theorem.

Suppose a strickly proper Gaussian field has the Markov property. If $U \subset D$ is open, we have $H(\overline{U}) \cap H(U') = H(\partial U)$ from the corollary to lemma 2. Let Π, P denote the projections onto $H(U')$, $H(\overline{U})$. From lemma 2, we have that Π, P commute and hence $\Pi + P - \Pi P$ is a projection. The range is clearly $H(\overline{U}) + H(U')$ and since the range of a projection is closed, the space $H(\overline{U}) + H(U')$ is closed.

By definition,

$H(\overline{U}) = \cap \{H(W): W \text{ is an open neighborhood of } \overline{U}\}$.

Now if $g \in S(D)$ has compact support, then $g \in H(W) + H(U')$, for there exists a C^∞ partition of unity α, β, for supp g subordinate to W and Int U'. Then $(\alpha + \beta)g = g$ and $\alpha g \in H(W)$, $\beta g \in H(U')$. Hence all C^∞ functions with compact support are contained in $H(W) + H(U')$. But since $H(\overline{U}) + H(U')$ is the intersection of such spaces, $H(\overline{U}) + H(U')$ contains this subspace. It follows immediately from the proof of lemma 1.7.2 in [3] that this subspace is dense in $S(D)$ and hence $H(\overline{U}) + H(U')$ is dense. Since it is closed, we have

$$H(\overline{U}) + H(U') = H(D).$$

Moreover, $H(U')^\perp \subset H(\overline{U})$, for, if F is orthogonal to $H(U')$

$$F = (\Pi + P - \Pi P)F = PF.$$

Suppose $f \in S(D)$ and supp $f \subset \overline{U}$. From lemma 3 we have Tf orthogonal to $H(W')$ for any open neighborhood W of $\overline{U}$. Hence, $Tf \in H(W)$. Choose a sequence $\{g_n\} \subset S(W)$ such that $\|Tf - g_n\|_R \to 0$. Then if $h \in S(D)$, supp $h \subset W'$,

$$0 = \langle g_n, h \rangle_2 = R(g_n, Th).$$

But since $\lim R(g_n, Th) = R(Tf, Th) = \langle Tf, h \rangle_2 = 0$, we have supp $Tf \subset \overline{W}$ and hence supp $Tf \subset \overline{U}$. Thus, T is local.

Correspondence Between Random and Quantum Fields.

In [7] Nelson investigated the generalized random Gaussian field over $S(R^n)$ whose covariance is $\langle (I - \Delta)^{-1} f, g \rangle$. Here, $(I - \Delta)^{-1}$ denotes the integral operator which solves a Poisson equation $u - \Delta u = f$, Δ is the Laplacian differential operator. Nelson established the Markov property and by considering the Markov condition for half spaces defined by one distinguished co-ordinate, x_1, Nelson constructed an appropriate operator family over $L^2(\phi, \mathcal{J}\{x \in R^n: x_1 = 0\})$ which is the operator system known as the free quantum field with mass one. Subsequently, Osterwalder and Schrader [8] showed that certain generalized functions defined by the inner products of a quantum field model satisfy properties similar to those of moments of random fields. A natural question is, do such generalized functions determine a random field, is the field Markov, and does the correspondence arise as with the free field? The following example, investigated by J. Challifour and myself, yields a negative answer for the last two questions.

The covariance form $\langle (I - \Delta)^{-\alpha} f, g \rangle_2$ is defined over $S(R^n)$ for any $0 < \alpha \leq 1$.
Using the Plancherel Theorem for the Fourier transform on $S(R^n)$, $f \to \hat{f}$, we have

$$\langle (I - \Delta)^{\alpha} f, g \rangle_2 = \langle (1 + |x|^2)^{\alpha} \hat{f}, \hat{g} \rangle_2 .$$

In fact, the right hand expression is usually taken as the defining expression for $(I - \Delta)^{\alpha} f$. It is clear that the form is strickly proper. However,

$T = (I - \Delta)^\alpha$ is not local, for if supp $f \subset \{x \in R^n : x_1 \leq 0\}$ then $\hat{f}$ has a unique continuous extension to $\mathrm{Im}\, x_1 > 0$ and the extension is analytic on $\mathrm{Im}\, x_1 > 0$ (see [3]). However, for such a test function,

$$(I - \Delta)^\alpha f^\wedge = (1 + |x|^2)^\alpha \hat{f}(x)$$

and if $f \neq 0$ then $\hat{f}(x) \neq 0$ for some $\mathrm{Im}\, x_1 > 0$. Then, $(I - \Delta)^\alpha f^\wedge$ can not be analytic for $\mathrm{Im}\, x_1 > 0$ and hence T is not local. From the theorem we conclude that the field is not Markov. The covariance form is generated by a quantum field model which is a superposition of free quantum fields with differing masses, m. The weighting measure for the mass is $(m^2 - 1)^{-\alpha}\, dm$, $m \geq 1$. The correspondence of these random and quantum fields may be verified by an involved computation using the methods of Osterwalder and Schrader.

References

[1] J.L. Doob, The elementary Gaussian processes. Annals. of Math. Stat. 15 (1944), 229-282.

[2] I.M. Gel'fand and N. Ya. Vilenkin, Generalized Functions, Vol. 4: Applications of Harmonic Analysis, Academic Press, N.Y., 1964.

[3] L. Hörmander, Linear Portial Differential Operators, Springer-Verlag, N.Y., 1969.

[4] H.P. McKean, Brownian motion with a several dimensional time. Theory Prob. Applications 8 (1963), 335-354.

[5] G.M. Molchan, On some problems concerning Brownian motion in Levy's sense. Theory Prob. Applications 12 (1967), 682-690.

[6] E. Nelson, Construction of quantum fields from Markoff fields, J. Functional Analysis 12 (1973), 97-112.

[7] E. Nelson, The free Markoff field, J. Functional Analysis 12 (1973), 221-227.

[8] K. Osterwalder and R. Schrader, Axioms for Euclidean Green's Functions, Commun. Math. Phys. 31, (1973), 83-113.

[9] L.D. Pitt, A Markov property for Gaussian processes with a multidimensional parameter, Arch. Rat. Mech. Anal. 43 (1971), 367-391.

A Survey of Abstract Wiener Spaces

by D. Kölzow

University of Erlangen-Nürnberg

I. Introduction

The purpose of Abstract Wiener Spaces is to describe Gaussian processes by pairs of spaces. The description is based on two ideas: a general one, applicable to many stochastic processes, and a special one, tailored to Gaussian processes. The general idea consists in describing a stochastic process by the following three data:
1. The path space W
2. the reproducing kernel Hilbert space H of the covariance
3. the embedding of H in W.

This idea leads to the following abstraction:

Definition: A **pair of spaces** is a pair (W,H), where:
1. W is a real locally convex Hausdorff space
2. H is a real Hilbert space
3. the following Axiom of Embedding is fulfilled.

Axiom of Embedding: H is a linear subspace of W such that the canonical embedding $i_{H,W}$ of H into W is continuous.

The special idea flows from the following consideration: By the embedding $i_{H,W}$ the canonical normal distribution on H is mapped to a Gaussian cylindrical measure $\mu_{W,H}$ on W. Now, the special idea is expressed by the

Axiom of Radonification: The cylindrical measure $\mu_{W,H}$ is extendable to a Radon measure $\tilde{\mu}_{W,H}$ on W. Here, a Radon measure on W means a regular Borel measure on the σ-algebra of Borelian subsets of W with respect to the weak topology on W.

Definition: An Abstract Wiener Space (AWS) is a pair of spaces (W,H) for which the Axiom of Embedding and the Axiom of Radonification are fulfilled. -Two AWS (W_1,H_1) and (W_2,H_2) are considered as equivalent and identified, if there is a linear homeomorphism of W_1 onto W_2 which maps H_1 isometrically onto H_2.

II. Examples of AWS

1. The classical AWS is a pair of spaces $(C_0[0,1]$, H), where $C_0[0,1]$ is the Banach space (with respect to the sup-norm) of all continuous real-valued functions on [0,1] vanishing at 0, and H is the Hilbert space of all those function $f \in C_0[0,1]$ which are absolutely continuous and differentiable such that $f' \in L_\mu^2[0,1]$, where μ is the Lebesgue measure on [0,1]. The scalar product of H is defined by $(f,g) := \int_0^1 f'g' d\mu$. In this case, $\tilde{\mu}_{W,H}$ turns out to be identical with the classical Wiener measure on $C_0[0,1]$. This fact may be taken as motivation of the name "Abstract Wiener Space". The classical AWS has been studied first by CAMERON-MARTIN [6],[7].

2. The pair $(\mathbb{R}^\mathbb{N}, l^2(\mathbb{N}))$ is an AWS, if $\mathbb{R}^\mathbb{N}$ is endowed with the topology of pointwise convergence,

and $L^2(\mathbb{N})$ is defined as usual. This AWS has been studied by SHILOV-FAN DYK TIN [35].

3. More closely related to the classical AWS than the latter ones are the AWS introduced by GROSS [13]. They are the pairs of spaces (W,H), where W is separable Banach space and H dense in W. The AWS (W,H) of GROSS can be described also by H and an additional continuous norm on H. In this way, GROSS originally defined them.

4. The most general AWS being considered up to now have been studied by DUDLEY-FELDMAN-LECAM [8]. They investigated AWS (W,H), where the corresponding measure $\mu := \widetilde{\mu}_{W,H}$ fulfills the following
<u>Tightness-Condition</u>: $\sup\{\mu(C): \begin{array}{l} C \subseteq W \text{ convex and} \\ \text{weakly compact} \end{array}\} = 1$

5. Finally, let H be a separable real Hilbert space, and a countably Hilbertian nuclear dense subspace of H (so that $\emptyset \subseteq H \subseteq \emptyset'$ is a Gelfand-Triplet, see GELFAND-VILENKIN [11]). Then, by the extension theorem of MINLOS [29], the pair of spaces $(\emptyset', H)$ is an AWS, where $\emptyset'$ is the topological dual of $\emptyset$. These AWS have been considered by GATTINGER [10], HIDA [17], HIDA-IKEDA [18], and UMEMURA [37].

III. Internal Structural Problems

1. Naturally, the Axiom of Radonification contains, in addition to the Axiom of Embedding, a further embedding condition.

<u>Problem 1</u>. Internal description of the Axiom of Radonification. A general solution of this problem is lacking. The following partial solutions are concerned with special pairs of spaces.

a) <u>The case W = H</u>: This case includes two subcases. The first one concerns the situation where the

topology of W is identical with the topology of H.
In this case (W,H) is an AWS iff dim H < ∞, as is
well-known. In the other situation, where the
topology of W is different from the topology of H,
only the special case where W is also a Hilbert
space has been investigated. SATO [32] proved that
in this case (W,H) is an AWS iff there is a non-
negative injective Hilbert-Schmidt operator on H
such that the norm $\| \ \|$ on H and the norm $\| \ \|_1$ on W
are related by $\|x_1\| = \|Ax\|$ for all $x \in H$.
This result of SATO can be reformulated as follows:
In case W is also a Hilbert space, (W,H) is an AWS
iff $i_{H,W}$ is a Hilbert-Schmidt operator.

b) <u>The case W Banach space, and H dense in W</u>: For
this case, first, GROSS [13] proved that (W,H) is an
AWS if the norm $\| \ \|_1$ of W is "measurable" with
respect to H in the following sense:

For every $\epsilon > 0$ there is a finite-dim.
projection P_ϵ on H such that for every finite-
dim. projection P on H orthogonal to P_ϵ we
have:

$$n\{h \in H: \|P(h)\|_1 > \epsilon\} < \epsilon,$$

where n denotes the canonical normal distribution on
H. Later, DUDLEY-FELDMAN-LECAM [8] proved that the
measurability of the norm of W with respect to H is
also necessary for (W,H) to be an AWS. In addition,
they gave two other characterizations of the
measurability of the norm of W with respect to H.
Further, SATO [32] proved that the norm of W can
be measurable with respect to H only if H is
separable. From this and from the results of
DUDLEY-FELDMAN-LECAM, mentioned above, it follows

that the separability of H (and so of B) is necessary for (W,H) to be an AWS.

To summarize, among the pairs of spaces (W,H) in question the AWS have been characterized by the norm of W. It would be interesting to get, in analogy to the reformulation of SATO's result for case a), a characterization by $i_{H,W}$.

2. Another way to get informations about the structure of AWS has been opened by SCHWARTZ [33], [34]. He investigated, for a given locally convex space W, the set Hilb(W) of all Hilbert spaces H such that (W,H) is a pair of spaces. For $H_1, H_2 \in$ Hilb(W) let $H_1 \approx H_2$ mean that H_1 and H_2 are isometrically isomorphic. Then, obviously: $H_1 \approx H_2 \Rightarrow \mu_{W,H_1} = \mu_{W,H_2}$. SCHWARTZ proved that the converse of this implication is also true. His proof is based on a factorization of the map $H \to \mu_{W,H}$ which may be sketched as follows:

$$\text{Hilb}(W) \to \mathfrak{L}^+(W) \to Q_c^+(W) \to \mathfrak{F}(W) \to \text{Cyl}(W),$$

where

$\mathfrak{L}^+(W)$ is the set of positive linear continuous maps $W' \to W$ (positive "kernels" with respect to W),

$Q_c^+(W)$ is the set of positive weak*-continuous quadratic forms $W' \to \mathbb{R}$,

$\mathfrak{F}(W)$ is the set of functions of positive type $F: W' \to \mathbb{R}$ such that $F(0) = 1$, and such that the restriction of F to every finite-dim. subspace of W' is continuous ("Fourier-transforms" with respect to W),

Cyl(W) is the set of cylindrical measures on W, and the factor maps are defined in the natural way.

IV. Identification of the Reproducing Kernel Hilbert Space of an AWS

Let (W,H) denote an AWS, $\mu := \tilde{\mu}_{W,H}$ the corresponding Gaussian Radon measure on W, and H(W) the reproducing kernel Hilbert space of the covariance of μ.

Theorem 1. $H \approx H(W)$ by the canonical embedding $H \subseteq W \subseteq R^{W'}$. This has been proved for the AWS of GROSS by KALLIANPUR [19], and for arbitrary AWS by GATTINGER [10].

For $w \in W$ let μ_w denote the translate of μ by w.

Theorem 2. $H = \{w \in W : \mu_w$ equivalent to $\mu\}$.
This has been proved by ROZANOV [31].

Because of Theorem 1, and according to the convention of identification for equivalent AWS, given in the Introduction, we can make the followign

Assumption. $H = H(W)$.

Theorem 3. supp $\mu = \overline{H}$

Theorem 4. A Zero-One-Law for AWS
If G is an additive subgroup of W, and $a \in W$ such that a + G is μ-measurable, then $\mu(a+G) \in \{0,1\}$.

Theorem 5. dim $H = +\infty \Rightarrow \mu(H) = 0$.

V. Construction of an AWS to a Gaussian Radon Measure

Let be given a real locally convex Hausdorff space W, and a zero Gaussian Radon measure μ on W. Let $H(\mu)$ denote the reproducing kernel Hilbert space of the covariance of μ.

Problem 2. Under which conditions exists a Hilbert space H such that (W,H) is an AWS and $\tilde{\mu}_{W,H} = \mu$ holds?

Theorem. If μ fulfills the Tightness-Condition, then there is a Hilbert space H such that (W,H) is an AWS, and $\tilde{\mu}_{W,H} = \mu$ holds. Further, H is dense in W iff μ fulfills the following

Condition (V): $\int \xi^2 d\mu > 0$ for every $\xi \in W'\setminus\{0\}$.

This was proved for the AWS of GROSS by SATO [32], and for arbitrary AWS by DUDLEY-FELDMAN-LECAM [8].

Corollary. If (W,H) is an AWS such that the Tightness-Condition is fulfilled for $\mu = \tilde{\mu}_{W,H}$ then: μ fulfills Condition (V) iff H is dense in W. This follows from Theorem 2 and the injectivity module $\approx$ of the map $H \to \mu_{W,H}$ (see Chap. III, Sect. 2).

Problem 3. Is the Tightness-Condition fulfilled for $\mu = \tilde{\mu}_{W,H}$ if (W,H) is an arbitrary AWS?

In case the answer to this problem is no, the following Problems arise:

Problem 4. Characterize Condition (V) by means of the embedding $i_{H,W}$.

Problem 5. Characterize the Condition $\overline{H} = W$ by means of the measure $\tilde{\mu}_{W,H}$.

VI. Measurable essentially linear (additive) Functionals on AWS

Let (W,H) denote an AWS, $\mu := \tilde{\mu}_{W,H}$ the corresponding Gaussian Radon measure on W, and H(W) the reproducing kernel Hilbert space of the covariance of μ. According to Chap. IV we can assume that H = H(W) holds.

<u>Definition</u>. A measurable essentially linear (additive) functional on (W,H) is a μ-measurable real-valued functional T on W for which a μ-measurable subspace $D = D(T)$ of W exists such that $\mu(D) = 1$ and the restriction of T on D is linear (additive).

<u>Lemma</u>. For every $h \in H$, regarded as an element of H', there is a μ-almost sure uniquely determined extension to a measurable essentially linear functional $\tilde{h}$ on (W,H).

This Lemma is basic for the following

<u>Representation-Theorem for measurable Functionals</u>

Let $(x_\alpha)_{\alpha \in A}$ denote an arbitrary orthonormal basis of H, and T a real-valued functional on W.

<u>Assertion</u>. T is measurable essentially linear (additive) on (W,H) iff there is a μ-almost uniquely determined $x_T \in H$ and a subspace X of W such that

1. X has outer μ-measure 1.
2. $T(x) = (x, x_T)^\sim$ for every $x \in X$, where

$$(x, x_T)^\sim := \sum_{\alpha \in A} T(x_\alpha) \, \tilde{x}_\alpha(X)$$

<u>Supplement</u>. If T is a measurable essentially linear (additive) functional on (W,H), then $H \subseteq D(T)$. This was proved for the AWS of GROSS by KUELBS [24], for $(R^N, l^2(N))$ by SHILOV-FAN DYK TIN [35], for AWS (W,H) with H separable by ROZANOV [31], and for arbitrary AWS by GATTINGER [10].

<u>Corollary</u> (Uniqueness-Theorem)

Two measurable essentially linear (additive) functionals on (W,H), coinciding on H, are identical.

The proof of this part of the Representation Theorem rests on the Zero-One Law of Chap. IV.

<u>Definition</u>: Let E denote a μ-measurable subspace of W such that $\mu(E) = 1$. A <u>kernel on E for (W,H)</u> is a map $K: \mathbb{R} \times W \to \mathbb{R}$ such that

1) $x \to K(t,x)$ is a measurable essentially linear functional on (W,H) with $D(K(t,\cdot)) = E$ for every $t \in \mathbb{R}$

2) $t \to K(t,x)$ is a continuous linear map for every $x \in E$

3) $x \to K(w'(x),x)$ is μ-integrable for every $w' \in W'$.

4) If $w'_n \to 0$ in the L^2_μ - norm, then $K(w'_n(\cdot),\cdot) \to 0$ in the L^1_μ - norm for every sequence (w'_n) in W'.

<u>Kernel-Representation Theorem</u>

Let (W,H) denote an AWS, and T a real-valued functional on W.

<u>Assertion</u>. T is μ-measurable essentially linear on (W,H) iff there is a kernel K on $D(T)$ for (W,H) such that

$$T(w') = \int_W K(w'(x), x) d_\mu(x) \text{ for every } w' \in W'.$$

The kernel K corresponding to a μ-measurable essentially linear Functional T on (W,H) is defined by

$$K(t,x) := t \cdot T(x).$$

Application of the Representation-Theorem for functionals yields the integral representation of T by K.

VII. Weakly measurable essentially linear Operators on AWS

Let (W,H) denote an AWS, and μ the corresponding measure.

Definition: A map on W in a real locally convex Hausdorff space is called <u>weakly measurable essentially linear on (W,H)</u>, if there is a μ-measurable subspace $D = D(T)$ of W such that

1. $\mu(D) = 1$
2. $F \cdot T$ is measurable and essentially linear on $D(T) \supseteq D$ on (W,H) for every $F \in V'$.

Remark. If $T: W \to V$ is weakly measurable essentially linear on (W,H), then $H \subseteq D(T)$.

Representation-Theorem for weakly measurable Transformations on $(\mathbb{R}^{\mathbb{N}}, l^2(\mathbb{N}))$.

A Transformation $T: \mathbb{R}^{\mathbb{N}} \to \mathbb{R}^{\mathbb{N}}$ is weakly measurable essentially linear on (W,H) iff there is a double-sequence $(a_{mn}) \in \mathbb{R}^{\mathbb{N} \times \mathbb{N}}$ such that

1. $\sum_{n=1}^{\infty} a_{mn}^2 < \infty$ for every $m \in \mathbb{N}$.
2. If $Tx = y$, where $x \in D(T)$, $x = (x_n)$, $y = (y_n)$, then $y_m = \sum_{n=1}^{\infty} a_{mn} x_n$ for μ-almost all $y \in \mathbb{R}^{\mathbb{N}}$.

This was proved by SHILOV-FAN DYK TIN [35] by means of the Representation-Theorem for measurable functionals (Chap. VI).

Representation-Theorem for weakly measurable Transformations on an AWS of GROSS.

Assumption. Let (W,H) be an AWS of GROSS, V a weakly sequential-complete separable Banach space, $T: W \to V$ a measurable essentially linear on (W,H), and (x_n) an orthonormal basis of H in W'.

Assertion. $T(x) = \sum_{n=1}^{\infty} T(x_n) \, x_n(x)$ for μ-almost all
$x \in W$.

This was proved by GATTINGER [10].

Problem 6. Extension of the Representation-Theorems for weakly measurable transformations to more general AWS.

VIII. Strongly measurable essentially linear Transformations on AWS

Let (W,H) denote an AWS, and μ the corresponding measure.

Definition: A transformation $T: W \to W$ is called **strongly measurable essentially linear on (W,H)**, if:
1. T is weakly measurable essentially linear on (W,H),
2. The image of μ under T is extendable to a Radon measure μ_T.
3. μ and μ_T are equivalent.

Theorem. If T is a strongly measurable essentially linear transformation on (W,H), then $T(H) \subseteq H$.

This was proved by FAN DYK TIN [35] and ROZANOV [31].

Representation Theorem for strongly measurable Transformations.

Assumption. Let H be separable and T be a weakly measurable essentially linear transformation on (W,H).

Assertion. T is strongly measurable iff I_H-$\text{Restr}_H T$ $(\text{Restr}_H T)^*$ is a Hilbert-Schmidt operator for which 1 is no eigen-value, where I_H denotes the identical map on H.

This has been proved by ROZANOV [31].

<u>Corollary</u>. <u>Assumption</u>. (W,H) be an AWS of GROSS and W be weakly sequentially complete. Further, let (x_n) be a complete orthonormal basis of H in W'.
<u>Assertion</u>. By $T \to (x_i, T(x_k))_{i,k \in \mathbb{N}}$ a bijection is defined, mapping the set of all strongly measurable essentially linear transformations on (W,H) onto the set of all real matrices $A = (a_{m,n})_{m,n \in \mathbb{N}}$ such that:
a) I-AA* is a Hilbert-Schmidt matrix for which 1 is no eigenvalue,
b) $\sum_{n=1}^{\infty} a_{mn}^2 < +\infty$ for all $m \in \mathbb{N}$, and $\sum_{m=1}^{\infty} a_{mn}^2 < +\infty$ for all $n \in \mathbb{N}$.

IX. Rotationally-invariant Measures on AWS

Let (W,H) denote an arbitrary AWS.
<u>Definition</u>. a) A continuous linear map $T: W \to W$ is called a <u>rotation on (W,H)</u>, if $x \in H \Leftrightarrow Tx \in H$ for every $x \in W$, and $\|Th\| = \|h\|$ for every $h \in H$ hold.
b) A cylindrical measure ν on W is called <u>rotationally-invariant on (W,H)</u>, if ν coincides with its image under any rotation on (W,H).
<u>Theorem</u>. Let (W,H) fulfill Condition (V).
<u>Assertion</u>. A cylindrical measure ν on W is rotationally-invariant on (W,H), iff there is a function $\varphi: \mathbb{R}_+ \to \mathbb{R}$ such that for the characteristic functional χ_ν of ν we have:

$$\chi_\nu(w') = \varphi(\|w'\|^2) \text{ for every } w' \in W'.$$

This has been proved by UMEMURA [37].
<u>Definition</u>: For every $t > 0$ let n_t denote the Gaussian distribution on H with mean zero and variance t, further, w_t denote the image of n_t under $i_{H,W}$ on W.

Lemma. If W is quasi-complete, w_t is extendable to a Radon measure $\tilde{w}_t$ on W.

This has been proved by BADRIKIAN-CHEVET [3].

Representation-Theorem for rotationally-invariant cylindrical Measures on AWS.

Assumption. (W,H) fulfill Condition (V), W be quasi-complete, and ν be a cylindrical measure on W.

Assertion. ν is rotationally-invariant on (W,H) iff there is a regular Borel measure ν' on R^+ such that

$$\nu(Z) = \int_{\mathbb{R}^+} \tilde{w}_t(Z) d\nu'(t)$$

for every cylinder set Z of W.

This has been proved by UMEMURA [37] for AWS associated with a Gelfand-Triplet (see Chap. II, Expl. 5), and by GATTINGER [10] as presented here.

X. Further Applications

a) **Tauberian Theorems for AWS.** Using the Representation Theorem for measurable Functionals (Chap. VI), KUELBS-MANDREKAR [26] extended the Tauberian Theorems of Wiener and Pitt to the AWS of GROSS. Later, GATTINGER [10] extended the Tauberian Theorem of Wiener to arbitrary AWS.

b) **Potential Theory on AWS** has been initiated by GROSS [16]. Harmonic functions on the AWS of GROSS have been studied also by GOODMAN [12], [42].

c) **Quasi-differentiable Functions** on separable Banach spaces have been introduced by GOODMAN [41], and investigated by means of an associated AWS of GROSS.

d) **A Gauss' Divergence Theorem** for the AWS of GROSS has been proved by GOODMAN [43].

e) <u>Parabolic Equations and Diffusion Processes</u> on the AWS of GROSS have been studied by PIECH [53-55], and KUO-PIECH [50].

f) <u>Stochastic Integrals</u> on the AWS of GROSS have been studied by KUO [47-50], and KUO-PIECH [51].

g) <u>Control Theory</u> on the AWS of GROSS has been introduced by KUO [52].

h) <u>White Noise</u> on AWS (W,H), where W is the dual of a nuclear space, has been investigated by HIDA [17], and HIDA-IKEDA [18].

i) <u>Strassen's Law of Iterated Logarithm</u> has been extended to the AWS of GROSS by KUELBS-Le PAGE [41], and to AWS (W,H), where W is the strict inductive limit of Frechet spaces, by KUELBS [42].

j) <u>Problems of linear Approximation and Estimation</u> for the AWS of GROSS have been studied by LARKIN [28].

k) <u>Problems of linear Prediction</u> can be restated for AWS. In this respect, the work of PARZEN [30] has to be mentioned. His Hilbert Space Representation of Time Series is closely related to the concept of AWS.

l) <u>Criteria for the Equivalence of two Gaussian Measures</u>, corresponding to two AWS (W,H_1) and (W,H_2), can be reformulated by means of H_1 and H_2 (cf. ROZANOV [31], and SKOROHOD [36]). This list of applications is certainly not complete.

XI. Final Questions

1. <u>How general is the Approach of GROSS?</u> The definition of AWS by pairs of spaces, as given in this paper, differs from the original approach of GROSS. While in our definition the relation between

the Hilbert space H and the path space W is
determined by the Axiom of Embedding, and the Axiom
of Radonification, GROSS is constructing W from H
by means of a measurable norm on H. As a consequence, if (W,H) is an AWS of GROSS, W is
necessarily a separable Banach space. But, the
approach of GROSS is not so restricted as it looks
at first glance, as the following two results show:
KUELBS [42] recently proved, that the Gaussian
measure corresponding to an AWS (W,H), where W is a
Frechet space, is supported by a separable Banach
subspace of W, provided the measure is tight.
KUELBS extended this result to the case, where W is
the strict inductive limit of Frechet spaces, under
certain additional assumptions of absolute continuity on the measure. But, KUELBS also showed by a
counter-example, that this result does not hold in
the case of an arbitrary complete locally convex
Hausdorff space W.
DUDLEY-FELDMAN-LECAM [8] succeeded in constructing
AWS (W,H) with Frechet spaces W by completing H by
means of a (countable) sequence of measurable norms.
But, still lacking is

Problem 7. Let (W,H) denote an arbitrary AWS. How
to reconstruct W by means of H?

Of course, this Problem 7 is closely related to
Problem 1, listed in Chap. III.

2. Where lie the historical Roots? To give an
exact answer to this question seems to be impossible.
Doubtlessly, WIENER [38], CAMERON, and MARTIN [6],
[7] initiated and investigated the classical AWS.
Concerning the reproducing kernel Hilbert space

of the covariance of a Gaussian process we mentioned already PARZEN [30], who in turns, refers to LOEVE [46]. On the other hand, integration of functionals on Hilbert spaces with respect to the canonical normal distribution has been developed by FRIEDRICHS and his collaborators [39], [40], as well as by SEGAL [56], [57].

3. <u>Does AWS Theory cover all Applications of Gaussian Measures?</u> This is definitely not the case for the AWS theory presented here. For example, let us call to mind that a main motivation for FRIEDRICHS and SEGAL to introduce an integration theory of such kind has been to solve problems in quantum field theory. For recent applications of Gaussian measures in this area see SIMON [58], and VELO-WIGHTMAN [59]. Here, certain limit processes for Gaussian measures are considered such that the limit measures are Markovian, and not Gaussian. This example indicates that the AWS theory, as presented here has to be extended or modified in order to cover all applications of Gaussian measures.

REFERENCES

1. N. ARONSZAJN, Theory of reproducing kernels. Trans. AMS 68 (1950), 337-404.
2. A BADRIKIAN, Séminaire sur les Fonctions Aléatoires Lineaires et les Mesures Cylindriques; Lecture Notes in Math. 139, Springer Verl. Berlin-Heidelberg-New York (1970).
3. A. BADRIKIAN/S. CHEVET, Questions liées à la théorie des espaces de Wiener; Colloque Internat. "Distributions aleatoires et les Processus gaussiens", Strasbourg, 1973.
4. A. BADRIKIAN/S. CHEVET, Mesures Cylindriques, Espaces de Wiener et Fonctions Aleatoires Gaussiennes; Lecture Notes in Math. 379, Springer Verl. Berlin-Heidelberg-New York (1970).
5. S. CAMBANIS/B. S. RAJPUT, Some zero-one laws for Gaussian processes; The Annals of Probability 1 (1973), 304-312.
6. R. H. CAMERON/W. T. MARTIN, Transformations of Wiener integrals under translations; Ann. of Math. (2) 45 (1944), 386-396; MR 6/5.
7. R. H. CAMERON/W. T. MARTIN, The Behaviour of Measure and Measurability under Change of Scale in Wiener Space; Bull. AMS 53 (1947), 130-137.
8. R. M. DUDLEY/J. FELDMAN/L. LE CAM, ON Seminorms and Probabilities, and Abstract Wiener Spaces; Ann. of Math. 93 (1971), 390-408.

9. FAN DYK TIN, Measurable linear functionals and transformations in a linear space with Gaussian measure; Uspeki Mat. Nauk 20 (1965) (3), 244-247.
10. M. GATTINGER, Representation theorems for measurable functionals and operators on Wiener spaces; Thesis, University of Erlangen-Nuremberg.
11. I. M. GELFAND/N. J. VILENKIN, Verallgemeinerte Funktionen IV; VEB Deutscher Verlag der Wissenschaften, Berlin (1964).
12. V. GOODMAN, A Liouville Theorem for Abstract Wiener Spaces; American J. Math. 45 (1973), 215-220.
13. L. GROSS, Measurable functions on Hilbert space; Trans. AMS 105 (1962), 372-390.
14. L. GROSS, Classical Analysis on a Hilbert space; in "Analysis in Function Space", Chap. IV ed. by W. T. Martin and I. Segal; The MIT Press, Massachusetts (1963).
15. L. GROSS, Abstract Wiener Spaces; Proc. of the Vth Berkeley Symp. on Math. Stat. and Prob. II (1) (1967), 31-42.
16. L. GROSS, Potential Theory on Hilbert Space; Journal Functional Analysis 1 (1967), 123-181.
17. T. HIDA, Stationary Stochastic Processes; Mathematical Notes, Princeton Univ. Press; Princeton, New Jersey (1970).
18. T. HIDA/N. IKEDA, Analysis on Hilbert Space with Reproducing Kernel Arising from Multiple Wiener Integral; Proc. of the Vth Berkeley Symp. on Math. Stat. and Prob., Berkeley-Los

Angeles (1967), II, 117-143.
19. G. KALLIANPUR, Zero-One Laws for Gaussian Processes; Trans. AMS <u>149</u> (1970), 199-211.
20. G. KALLIANPUR, The role of reproducing kernel Hilbert space in the study of Gaussian processes; in: "Advances in Probability and Related Topics" <u>II</u>, ed. by P. Ney, Marcel Dekker Inc., New York (1970), 49-83.
21. G. KALLIANPUR, Abstract Wiener processes and their reproducing kernel Hilbert spaces; Zeitschr. für Wahrsch. theorie <u>17</u> (1971), 113-123.
22. G. KALLIANPUR/M. NADKARNI, Supports of Gaussian Measures; Proc. of the VIth Berkeley Symp. on Math. Stat. and Prob., <u>II</u> (1970), 375-387; University of California Press, Berkeley and Los Angeles (1972).
23. D. KÖLZOW, Functional integration, part I; Cylindrical measures on abstract vector spaces; Lecture Notes, University of Florida (1969/70).
24. J. KUELBS, Abstract Spaces and Applications; Pacific J. of Math. <u>31</u> (2) (1969), 433-450.
25. J. KUELBS, Gaussian measures on a Banach space; J. Functional Analysis <u>5</u> (1970), 345-367.
26. J. KUELBS/B. MANDREKAR, The Wiener Closure Theorem for Abstract Wiener Spaces; Proc. AMS <u>32</u> (1972), 169-178.
27. J. KUELBS/F. M. LARKIN/J. A. WILLIAMSON, Weak probability distributions on reproducing kernel Hilbert spaces; Rocky Mountain J. of Math. <u>2</u> (1972), 369-378.

28. F. M. LARKIN, Gaussian measure in Hilbert space and appl. in numerical analysis; Rocky Mountain J. of Math. <u>2</u> (1972), 379-421.
29. R. A. MINLOS, Generalized random processes and their extension to measures; Trudy Moskov. Mat. Obsc. <u>8</u> (1959), 497-518.
30. E. PARZEN, Statistical inference on time series by RKHS methods; Dept. of Statistics, Technical Report # 14, Stanfor Univ., Stanford Calif., (1970).
31. YU. A. ROZANOV, Infinite-dimensional Gaussian Distributions; Proc. of the Steklov Institute of Math. <u>108</u> (1968), Izdat. "Nauka", Moskau; American Math. Soc., Providence Rhode Island (1971).
32. H. SATO, Gaussian measures on a Banach Space and abstract Wiener Space; Nagoya Math J. <u>36</u> (1969), 65-81.
33. L. SCHWARTZ, Sous-espaces hilbertiens d'espaces vectoriels topologiques et noyaux associes. (Noyaux reproduisants.) J. Analyse Math. <u>13</u> (1964), 115-256.
34. L. SCHWARTZ, Radon Measures on Arbitrary Topological Spaces. Preprint.
35. G. E. SHILOV/FAN DYK TIN, Integral, Measure and Derivative on Linear Spaces; Izdat. "Nauka", Moskau (1967), MR 37/1554.
36. A. V. SKOROHOD, Integration in Hilbert Space. Ergebnisse der Mathematik und ihrer Grenzgebiete. Bd. 79. Springer-Verl. Berlin (1974), 177 S.

37. Y. UMEMURA, Measures on infinite dimensional vector spaces; Publications of the Research Inst. for Math. Sciences of Kyoto University $\underline{1}$ (1966), 1-47.
38. N. WIENER, The Average Value of a Functional. Proc. London Math. Soc. $\underline{22}$ (1924), 454-467.

ADDITIONAL REFERENCES

39. K. O. FRIEDRICHS, Mathematical aspects of the quantum theory of fields; Interscience Pub., New York and London (1953).
40. K. O. FRIEDRICHS/H. N. SHAPIRO et al., Integration of functionals; Seminar Notes. New York University, Inst. of Math. Sci. (1957)
41. V. GOODMAN, Quasi-differentiable functions on Banach spaces; Proc. AMS $\underline{30}$ (1971), 367-370.
42. V. GOODMAN, Harmonic functions on Hilbert space; Journal Functional Analysis $\underline{10}$ (1972), 451-470.
43. V. GOODMAN, A divergence theorem for Hilbert space Trans AMS $\underline{164}$ (1972), 411-426.
44. J. KUELBS/R. LE PAGE, The law of iterated logarithm for Brownian motion in a Banach space; Trans. AMS (to appear).
45. J. KUELBS, Some results for probability measures on linear topological vector spaces with an application to Strassen's log log law; Journal Functional Analysis $\underline{14}$ (1973), 28-43.
46. M. LOEVE, Fonctions aléatoires du second ordre; Supplement to P. LEVY: Processus Stochastiques et Mouvement Brownien; Paris, Gauthier-Villars (1948).

47. H. H. KUO, Stochastic integrals in abstract Wiener space. Pacific Journal Math. 41 (1972), 459-469.
48. H. H. KUO, Stochastic integrals in abstract Wiener space II: Regularity properties; Nagoya Math. J. (to appear).
49. H. H. KUO, On operator-valued stochastic integrals; Bull. AMS 79 (1973), 207-210.
50. H. H. KUO/M. A. PIECH, Stochastic integrals and parabolic equations in abstract Wiener space; Bull. AMS 79 (1973), 478-482.
51. H. H. KUO, Differential and Stochastic equations in abstract Wiener space; Journal Functional Analysis 12 (1973), 246-256.
52. H. H. KUO, On a stochastic maximum principle in Banach space; Journal Functional Analysis 14 (1973), 146-161.
53. M. A. PIECH, A fundamental solution of the parabolic equation on Hilbert space. I: Journal Functional Analysis 3 (1969), 85-114. II: Trans. AMS 150 (1970), 257-286.
54. M. A. PIECH, Some regularity properties of diffusion processes on abstract Wiener spaces; Journal Functional Analysis 8 (1971), 153-172.
55. M. A. PIECH, Diffusion semigroups on abstract Wiener space. Trans. AMS 166 (1972), 411-430.
56. I. SEGAL, Non-commutative extension of abstract integration; Ann. Math. 57 (1953), 77-114.
57. I. SEGAL, Distribution in Hilbert space and canonical systems of operators; Trans. AMS 88 (1958), 12-41.

58. B. SIMON, The $P(\phi)_2$ Euclidean (quantum) field theory. Princeton Series in Physics; Princeton University Press (1974).
59. G. VELO/A. WIGHTMAN, ed., Constructive quantum field theory. The 1973 "Ettore Majorana" International School of Mathematical Physics. Lecture Notes in Physics $\underline{25}$ (1973), Springer-Verlag Berlin, Heidelberg, New York.

QA
274
S964
1974
v.1

JAN 7 1976